Benzimidazole

Edited by Pravin Kendrekar
and Vinayak Adimule

Published in London, United Kingdom

Benzimidazole
http://dx.doi.org/10.5772/intechopen.95148
Edited by Pravin Kendrekar and Vinayak Adimule

Contributors
Thatikayala Mahender, Anil Kumar Garige, Hemalatha Gadegoni, Mohammad Shahar Yar, Kashif Haider, Vinayak Adimule, Pravin Kendrekar, Sheetal Batakurki, Akhil Sivan, Gopakumar Kavya, A. Antony Muthu Prabhu, Matloob Ahmad, Imran Ahmad Khan, Noor ul Amin Mohsin, Sana Aslam

First published in London, United Kingdom, 2022 by IntechOpen
IntechOpen is the global imprint of INTECHOPEN LIMITED, registered in England and Wales, registration number: 11086078, 5 Princes Gate Court, London, SW7 2QJ, United Kingdom
Printed in Croatia

British Library Cataloguing-in-Publication Data
A catalogue record for this book is available from the British Library

Additional hard and PDF copies can be obtained from orders@intechopen.com

Benzimidazole
Edited by Pravin Kendrekar and Vinayak Adimule
p. cm.

This title is part of the Biochemistry Book Series, Volume 34
Topic: Chemical Biology
Series Editor: Miroslav Blumenberg
Topic Editors: Şükrü Beydemir and Deniz Ekinci

Print ISBN 978-1-80355-120-3
Online ISBN 978-1-80355-121-0
eBook (PDF) ISBN 978-1-80355-122-7
ISSN 2632-0983

For EU product safety concerns:
IN TECH d.o.o., Prolaz Marije Krucifikse Kozulić 3, 51000 Rijeka, Croatia,
info@intechopen.com or visit our website at intechopen.com.

IntechOpen Book Series
Biochemistry
Volume 34

Aims and Scope of the Series

Biochemistry, the study of chemical transformations occurring within living organisms, impacts all of the life sciences, from molecular crystallography and genetics, to ecology, medicine and population biology. Biochemistry studies macromolecules - proteins, nucleic acids, carbohydrates and lipids –their building blocks, structures, functions and interactions. Much of biochemistry is devoted to enzymes, proteins that catalyze chemical reactions, enzyme structures, mechanisms of action and their roles within cells. Biochemistry also studies small signaling molecules, coenzymes, inhibitors, vitamins and hormones, which play roles in the life process. Biochemical experimentation, besides coopting the methods of classical chemistry, e.g., chromatography, adopted new techniques, e.g., X-ray diffraction, electron microscopy, NMR, radioisotopes, and developed sophisticated microbial genetic tools, e.g., auxotroph mutants and their revertants, fermentation, etc. More recently, biochemistry embraced the 'big data' omics systems. Initial biochemical studies have been exclusively analytic: dissecting, purifying and examining individual components of a biological system; in exemplary words of Efraim Racker, (1913 –1991) "Don't waste clean thinking on dirty enzymes." Today, however, biochemistry is becoming more agglomerative and comprehensive, setting out to integrate and describe fully a particular biological system. The 'big data' metabolomics can define the complement of small molecules, e.g., in a soil or biofilm sample; proteomics can distinguish all the proteins comprising e.g., serum; metagenomics can identify all the genes in a complex environment e.g., the bovine rumen.
This Biochemistry Series will address both the current research on biomolecules, and the emerging trends with great promise.

Meet the Series Editor

Miroslav Blumenberg, Ph.D., was born in Subotica and received his BSc in Belgrade, Yugoslavia. He completed his Ph.D. at MIT in Organic Chemistry; he followed up his Ph.D. with two postdoctoral study periods at Stanford University. Since 1983, he has been a faculty member of the RO Perelman Department of Dermatology, NYU School of Medicine, where he is codirector of a training grant in cutaneous biology. Dr. Blumenberg's research is focused on the epidermis, expression of keratin genes, transcription profiling, keratinocyte differentiation, inflammatory diseases and cancers, and most recently the effects of the microbiome on the skin. He has published more than 100 peer-reviewed research articles and graduated numerous Ph.D. and postdoctoral students.

Meet the Volume Editors

Dr. Pravin Kendrekar, MSc, MBA, Ph.D., is currently a visiting scientist at the Lipid Nanostructure Laboratory, University of Central Lancashire, England. He previously worked as a post-doctoral fellow at the Ben-Gurion University of Negev, Israel; University of the Free State, South Africa; and Central University of Technology Bloemfontein, South Africa. He obtained his Ph.D. in Organic Chemistry from Nagaoka University of Technology, Japan. He has published more than seventy-four journal articles and attended several national and international conferences as speaker and chair. Dr. Kendrekar has received many international awards. He has several funded projects, namely, anti-malaria drug development, MRSA, and SARS-CoV-2 activity of curcumin and its formulations. He has filed four patents in collaboration with the University of Central Lancashire and Mayo Clinic Infectious Diseases. His present research includes organic synthesis, drug discovery and development, biochemistry, nanoscience, and nanotechnology.

Dr. Vinayak Adimule, MSc, Ph.D., is a professor and dean of R&D, Angadi Institute of Technology and Management, India. He has 15 years of research experience as a senior research scientist and associate research scientist in R&D organizations. He has published more than fifty research articles as well as several book chapters. He has two Indian patents and two international patents to his credit. Dr. Adimule has attended, chaired, and presented papers at national and international conferences. He is a guest editor for *Topics in Catalysis* and other journals. He is also an editorial board member, life member, and associate member for many international societies and research institutions. His research interests include nanoelectronics, material chemistry, artificial intelligence, sensors and actuators, bio-nanomaterials, and medicinal chemistry.

Contents

Preface

Benzimidazole is an aromatic, heterocyclic molecule containing C, H, N, and S formed from the benzene and imidazole moieties. Benzimidazole derivatives have attracted much interest from researchers due to their biological activities and clinical applications. Benzimidazole nucleus is also a constituent of Vitamin B12. Benzimidazole derivatives act as promising bioactive compounds and exhibit a range of biological activities like anticancer, antimicrobial, antiviral, anticonvulsant, antiproliferative, antioxidant, and antiparasitic properties. The increased interest in benzimidazole compounds is due to their bioavailability, increased stability, significant biological activity, and so on.

A vast number of benzimidazole derivatives have been reported and studies on their mechanism of action, structural features, and pharmacological applications have been carried out to achieve more stable and active pharmaceutical drugs. Interesting features of benzimidazole derivatives include their N-donor ligands and the physical interaction of double bound with other ring systems.

This book examines aspects and newer mechanisms of benzimidazoles containing heterocyclic moiety. Chapters report on anticancer properties of benzimidazole derivatives, novel methods of synthesis of benzimidazoles, versatile nature of the benzimidazoles, spectral and theoretical studies of benzimidazole derivatives, and medicinal importance and pharmacological aspects of benzimidazole derivatives.

The editors are thankful to Lancashire University, UK, and the director of the Angadi Institute of Technology and Management for her support.

Dr. Pravin Kendrekar
Visiting Scientist,
Lipid Nanostructure Laboratory,
University of Central Lancashire, Preston,
Preston, United Kingdom

Dr. Vinayak Adimule
Professor and Dean R&D,
Department of Chemistry,
Angadi Institute of Technology and Management,
Belagavi, India

Section 1

Anticancer Properties of Benzimidazole

Chapter 1

Advances of Benzimidazole Derivatives as Anticancer Agents: Bench to Bedside

Kashif Haider and Mohammad Shahar Yar

Abstract

Benzimidazole is one of the privileged nitrogen-containing scaffolds known for its versatile diversified role in insecticides, pesticides, dyes, pigments and pharmaceuticals. Due to its electron-rich environment, structural features and binding potency of various therapeutic targets, benzimidazole derivatives exhibit a broad spectrum of biological activity that majorly includes antimicrobial, antifungal, analgesics, anti-diabetic and anticancer agents. Several benzimidazole scaffolds bearing drugs are clinically approved; they are used for various indications. For example, Bilastine, Lerisetron, Maribavir and Nocodazole are the most widely used benzimidazole-based marketed drugs available as an antihistamine, antiviral and antimitotic agent, respectively. Another example is the recently approved anticancer drug Binimetinib and Selumetinib, which are indicated for BRAF mutated melanoma and plexiform neurofibromas. Not only this, many benzimidazole-based anticancer drugs are in late phases of clinical development. Due to the vast therapeutic potential of benzimidazole scaffold in cancer research, medicinal chemists have gained a lot of attraction to explore it more and develop novel, highly effective and target-specific benzimidazole-based potential anticancer drugs.

Keywords: benzimidazole, enzyme inhibitors, anticancer agents, hybrid derivatives

1. Introduction

Cancer is a complex, severe class of diseases that involves a group of cells that exhibit abnormal and uncontrolled division and proliferation. It is one of the primary health concerns which accounts for the second major cause of death globally. As per the recent statistics of the world health organization (WHO), in 2020, around 10 million people succumbed to death due to cancer. However, every year the number of incidences is increasing day by day. According to WHO, around 0.3 million new cases are diagnosed each year among the age group of 0–19 years. Cancer can affect a person of any age; however, with age, the risk increases. Globally, steady increases in cancer cases every year are taking a toll on the health care system [1–5]. To combat cancer, identification of potential drugs and potential drugs combination is essential. Potential research has been carried out to counter such problems by addressing novel drug design and discovery approaches. In medicinal chemistry, heterocyclic rings have played a significant role in the search for potential therapeutic agents. Various drugs are currently in use and in development that widely addresses

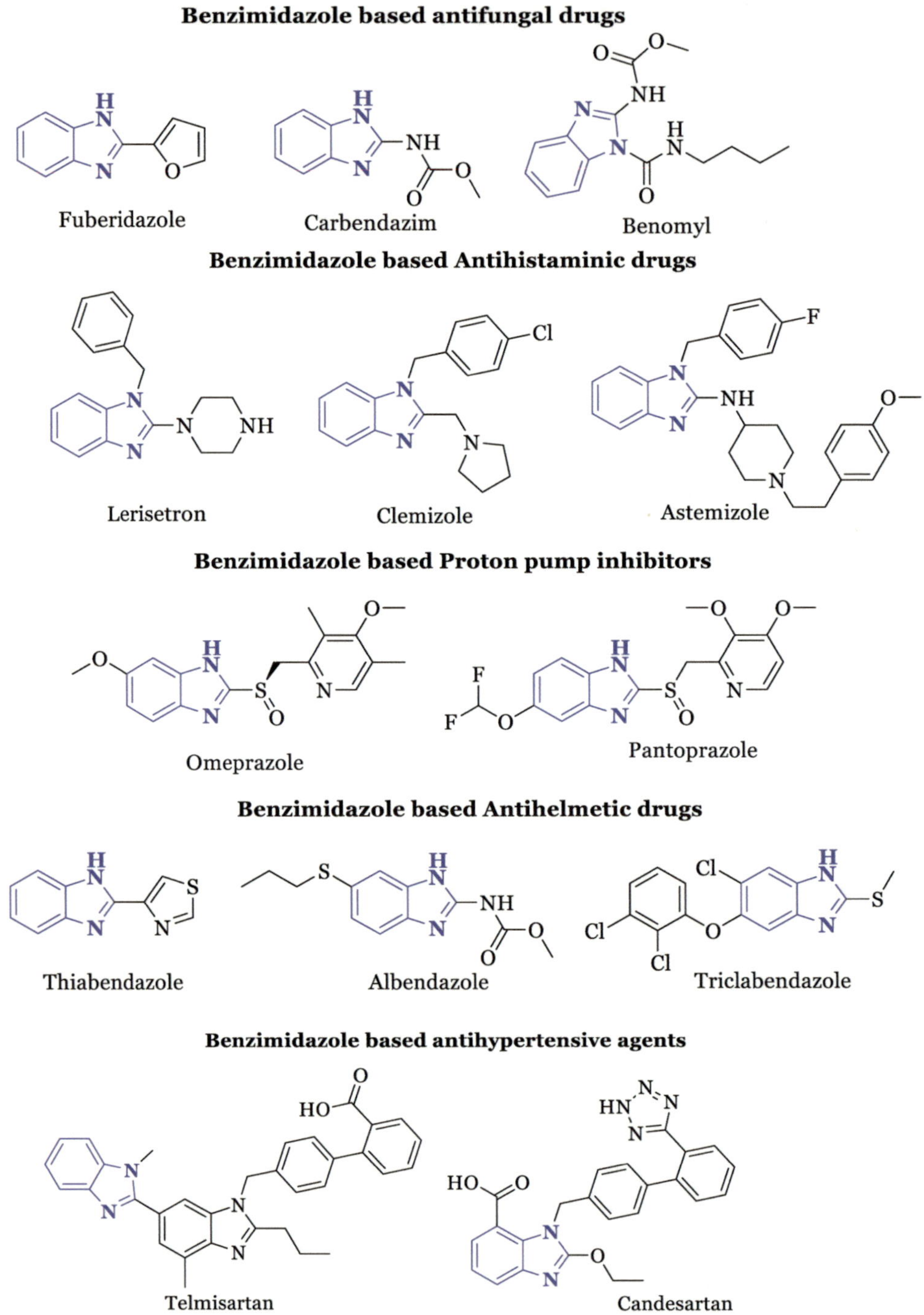

Figure 1.
Examples of benzimidazole based drugs in clinical use.

such problems. However, due to changes in cancer forms and mutations, current therapy faces challenges of poor selectivity and specificity towards certain types of cancer cells, which narrows down their effectiveness. Generally, cancer cells act by disrupting and disturbing the cell signaling pathways; therefore, it is crucial to design novel target-based heterocyclic anticancer compounds with high efficacy and fewer side effects, which will provide a solid backup to the present chemotherapeutic regime [6–10].

2. Benzimidazole

Benzimidazole is a bicyclic nitrogen bearing aromatic heterocyclic ring, structurally it consists of benzene ring fused with imidazole ring at the 4th and 5th position of the ring. Chemically it appears as white crystals, amphoteric in nature, resembles the structure of purine. It is synthesized by different reported methods. However, condensation of 1,2-diamino benzene with carbonyl compounds to give benzimidazole is the conventional method which was used widely for its preparation. In 1858, it was synthesized by Heinrich Debus, a German chemist from glyoxal, ammonia and formaldehyde, that's why it was also known as glyoxalin. Benzimidazole ring is one of bioactive heterocyclic scaffold exhibiting wide range of biological activities. The ■NH group present at second position of the ring is both highly acidic and weak base in nature, it also has ability to form stable salts [11–16].

With time benzimidazole ring emerged as an important multifaceted heterocyclic system due to its wide range of pharmacological activity such as antibacterial [17], antiparasitic [18], antifungal [19], anti-inflammatory [20], analgesics [21], antiviral [22], antitubercular [23], anticoagulant [24], antihistaminic [25], antioxidant [26], antiulcer [27] and anticancer [28–31]. Some of the benzimidazole based marketed drugs are listed in with their indication and marketed name in **Figure 1**. Adding to this benzimidazole scaffold have also displayed a significant role in synthesis of organic intermediates. In light of the application of benzimidazole earlier various authors have reported many review articles. Due to the diverse therapeutic potential, benzimidazole have attracted lot of researchers to explore more in the field of drug discovery to synthesize novel and potent compounds with a broad spectrum of biological activities. Owing to this, with time efforts have been made to create libraries of these potent compounds. In cancer treatment benzimidazole based drugs played a significant role, various targeted therapies are designed and developed as Kinase inhibitors such as EGFR, VEGFR and PI3K inhibitors here, in this chapter we have included some potent benzimidazole based kinase inhibitors.

3. Advances of benzimidazole based anticancer agents

Benzimidazole based compounds have got much attention due to exhibiting significant cytotoxic activity. In last one decade a lot of benzimidazole based anticancer drugs have received status of US FDA global approval. Recently, Binimetinib, Selumetinib and Abemaciclib got approval for treatment of various mutated forms of cancer. Here, we have discussed some of benzimidazole based anticancer drugs which are recently approved, under development and in pipeline.

3.1 Benzimidazole based marketed anticancer drugs

*3.1.1 Binimetinib (**1**)*

Binimetinib (**1**) is chemically 5-((4-bromo-2-fluorophenyl)amino)-4-fluoro-N-(2-hydroxyethoxy)-1-methyl-1H-benzimidazole-6-carboxamide approved by US FDA recently in 2018. It is an orally available, potent selective inhibitor of mitogen activated protein kinase (MEK 1/2). Binimetinib is developed by Array Biopharma, commercially available by the name of Mektovi. It is indicated for patients having metastatic melanoma with BRAF mutation as combination therapy with BRAF inhibitors Encorafenib [32]. Presently, Binimetinib is in various phases of clinical development as monotherapy or in combination for conditions like KRAS mutated cancer, mutated non-small cell lung cancer [33, 34]. Structures of all the drugs are presented in **Figure 2**. More details of clinical trials are enlisted in **Table 1**.

*3.1.2 Bendamustine (**2**)*

Bendamustine (**2**) is chemically 4-(5-(bis(2-chloroethyl)amino)-1-methyl-1H-benzimidazol-2-yl)butanoic acid, it is an alkylating agent well known for its efficacy and tolerability in wide range of hematologic malignancies [35]. Bendamustine is indicated for the treatment of chronic lymphocytic leukemia and non-Hodgkin lymphoma [36]. Currently Bendamustine is further investigation as combination therapy along with Bcl-2 inhibitor Venetoclax and Rituximab for treatment of patient above 60 years of age with mantle cell lymphoma (NCT03834688).

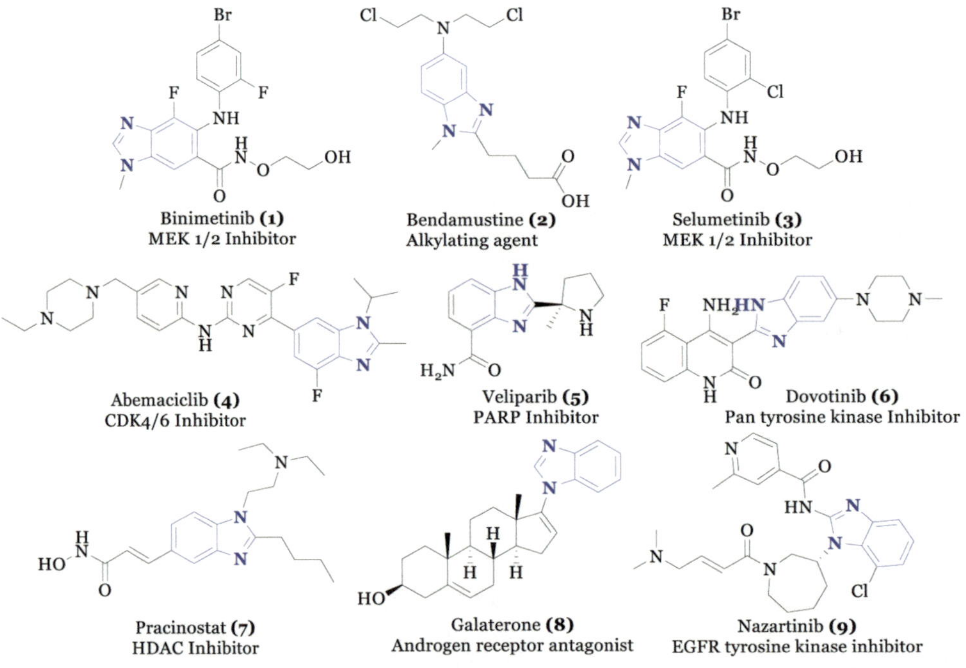

Figure 2.
Benzimidazole based clinically approved anticancer agents.

Drug	Clinical trial number	Clinical trial study	Date of study	Current status and study phase
Binimetinib	NCT04965818	Phase 1b/2 study of Futibatinib in combination with Binimetinib in patients with advanced KRAS mutant cancer	Last update on September 27, 2021	Recruiting Phase 1b/2
	NCT03170206	Study of CDK4/6 inhibitor Palbociclib in combination with the Binimetinib for patients with advanced KRAS mutant NSCLC	Last update on June 10, 2021	Recruiting Phase 1
Bendamustine	NCT04217317	CPI-613 in combination with Bendamustine in patients with relapsed or refractory T-cell Non-Hodgkin lymphoma	Last update on August 30,2021	Recruiting Phase 2
	NCT04510636	Study of Pembrolizumab with Bendamustine in Hodgkin lymphoma	Last update on August 30,2021	Not yet Recruiting Phase 2
Selumetinib	NCT02768766	Intermittent Selumetinib for uveal melanoma	Last update on March 19, 2021	Recruiting Phase 1
	NCT05101148	Phase I study to assess the effect of food on the PK and gastrointestinal toxicity of Selumetinib in adolescent children with Neurofibromatosis Type 1 related plexiform neurofibromas	Last update on November 1, 2021	Recruiting Phase 1
Abemaciclib	NCT04003896	A study to evaluate Abemaciclib in advanced biliary tract carcinoma who failed prior first line therapy.	Last update on	Active, Not recruiting Phase 2
	NCT04040205	Abemaciclib for bone and soft tissue sarcoma with cyclin dependent kinase (CDK) pathway attention	February 15, 2021	Recruiting Phase 2
Veliparib	NCT02723864	Veliparib and VX-970 in combination with cisplatin in people with refractory solid tumors	Last update on February 5, 2021	Active, Not recruiting Phase 1
	NCT01434316	Veliparib and Dinaciclib in treating patients with advanced solid tumors	July 20, 2021	Recruiting Phase 1
Dovitinib	NCT01635907	Dovitinib in neuroendocrine tumors	Last update on April 14, 2020	Completed Phase 2

Drug	Clinical trial number	Clinical trial study	Date of study	Current status and study phase
Pracinostat	NCT03848754	Pracinostat and Gemtuzumab ozogamicin in patients with relapsed or refractory acute myeloid leukemia	Last update on October 18, 2021	Active, not recruiting Phase 1
Galeterone	NCT04098081	Galeterone with Gemcitabine for patients with metastatic pancreatic adenocarcinoma	Last update on March 10, 2021	Recruiting Phase 2
Nazartinib	NCT02335944	Study and safety and efficacy of Nazartinib in combination with cMET inhibitor INC280 in NSCLC patients with EGFR mutation	Last update on October 4, 2021	Active, not recruiting Phase 1/2
	NCT02108964	A phase I/II, multicentre, open label study of Nazartinib, administered orally in adult patients with EGFR mutated solid malignancies	Last update on August 13, 2021	Active, Not recruiting Phase 1/2

Table 1.
Benzimidazole based anticancer drugs in clinical development.

*3.1.3 Selumetinib (**3**)*

Selumetinib (**3**) is chemically 5-((4-bromo-2-chlorophenyl)amino)-4-fluoro-N-(2-hydroxyethoxy)-1-methyl-1H-benzimidazole-6-carboxamide recently approved by US FDA in on April 10, 2020 for the treatment of plexiform neurofibromas and neurofibromatosis in pediatric patients [37, 38]. Selumetinib is an orally available MEK 1/2 kinase inhibitor developed by AstraZeneca commercially available by the name of Koselugo. It is also received status of orphan drug in USA as adjuvant drug for treatment of thyroid cancer [39, 40].

*3.1.4 Abemaciclib (**4**)*

Abemaciclib (**4**) is chemically N-(5-((4-ethylpiperazin-1-yl)methyl)pyridin-2-yl)-5-fluoro-4-(4-fluoro-1-isopropyl-2-methyl-1H-benzimidazol-6-yl) pyrimidin-2-amine, approved by US FDA on 28 September 2017, for the treatment of patients with hormone receptor (HR) positive, human epidermal growth factor receptor (HER-2) advanced/negative metastatic breast cancer as a combination therapy with estrogen receptor antagonist fulvestrant in female patients and as monotherapy in adult patient with diseases progression following chemotherapy. Abemaciclib is commercially available by the name of Verzenio, developed by Eli Lilly as cyclin dependent kinase-4 (CDK4) and CDK6 inhibitor [41]. Furthermore, Abemaciclib is currently in various phase of clinical development as monotherapy or in combination therapy for treatment of various types of cancer and mutated forms [42, 43].

*3.1.5 Veliparib (**5**)*

Veliparib (**5**) is chemically (R)-2-(2-methylpyrrolidin-2-yl)-1H-benzimidazole-4-carboxamide, it is an oral PARP inhibitor. Veliparib is investigational drug showed promising results in preclinical and clinical studies when treated for ovarian cancer and for mutated form BRCA-mutated ovarian cancer [44]. Further development of Veliparib is ongoing as monotherapy and combination therapy for treatment of different forms of ovarian cancer [45, 46].

*3.1.6 Dovitinib (**6**)*

Dovitinib (**6**) is chemically 4-amino-5-fluoro-3-(5-(4-methylpiperazin-1-yl)-1H-benzoimidazol-2-yl) quinolin-2(1H)-one, it is a potent orally available pan tyrosine kinase inhibitor targeting VEGFR, FGFR) and other tyrosine kinases [47]. It is a pipeline drug under development, for treatment of gastrointestinal stromal tumor, metastatic breast cancer and renal cell carcinomas. Recently on April 2, 2021 Dovitinib has received acceptance from US FDA for premarket approval (PMA) which was filed by Allarity therapeutics (details can be found on Allarity therapeutics website). Dovitinib is also explored for different typed of mutated forms of cancer, currently it is under phase II clinical trial study for patient with castration resistant prostate cancer [48].

3.1.7 Pracinostat (7)

Pracinostat (7) is chemically (E)-3-(2-butyl-1-(2-(diethyl amino) ethyl)-1H-benzoimidazol-5-yl)-N-hydroxyacrylamide, it is orally available, investigational drug exhibiting potential antitumor activity [49, 50]. Pracinostat is a small molecule next generation histone diacetylases (HDAC) inhibitor indicated acute myeloid leukemia [51]. In some recent study Pracinostat was found to suppresses growth and metastasis of breast cancer by inactivating the IL-6/STAT3 signaling pathway [52].

*3.1.8 Galeterone (**8**)*

Galeterone (**8**) is chemically (3S,8R,9S,10R,13S,14S)-17-(1H-benzimidazol-1-yl)-10,13-dimethyl-2,3,4,7,8,9,10,11,12,13,14,15-dodecahydro-1H-cyclopenta[a]phenanthren-3-ol, it is an orally available, small molecule investigational drug. Galeterone is developed by Tokai pharmaceutical as potent androgen receptor antagonist, indicated for treatment of prostate cancer [53]. Some in vivo studies revealed that Galeterone monotherapy inhibited breast cancer growth, also when administered in combination with cisplatin the results where promising and much better compare to monotherapy of cisplatin [54].

*3.1.9 Nazartinib (**9**)*

Nazartinib (**9**) is chemically (R,E)-N-(7-chloro-1-(1-(4-(dimethylamino) but-2-enoyl)azepan-3-yl)-1H-benzo[d]imidazol-2-yl)-2-methylisonicotinamide, it is an orally available third generation EGFR kinase inhibitor under development for treatment of conditions like non-small cell lung cancer (NSCLC) and EGFR mutated NSCLC [55]. Nazartinib have demonstrated favorable safety profile and efficacy in a Phase-I study when administered to adult patients with

EGFR mutated lung carcinoma [56]. However, clinical development of Nazartinib is progress for different forms of mutated carcinomas as monotherapy or in combination [57, 58].

3.2 Benzimidazole based derivatives as potent kinase inhibitors

Commonly the mechanism behind action of anticancer agents involve DNA intercalation, gene regulation, microtubule inhibition, transcription regulation, DNA synthesis inhibition, enzyme inhibition and so on. Nowadays in cancer treatment, target therapy emerged as one of the acknowledged strategies. Most of the available anticancer drugs acts by targeting structural proteins, tyrosine kinases, phosphoinositide 3 kinase and protein kinases for example Binimetinib acts by inhibiting mitogen activated kinase as discussed in earlier section. In this section we have included some recent examples of benzimidazole based enzyme inhibitors as potent anticancer agents.

3.2.1 EGFR inhibitors

Akhter et al. have reported a novel series of benzimidazole based oxadiazole derivatives as potential EGFR inhibitors. The target compound **10** and **11** demonstrated significant binding to EGFR with an IC_{50} value of 0.081 and 0.098 μM respectively. Cytotoxicity of both derivatives against selected human cancer cell line A549, MDA-MB231, MCF7 and HepG2 was found promising. Compound **10** exhibited excellent inhibitory potency with an IC_{50} value of 15.2 μM, 5.0 μM, 14.5 μM and 12.5 μM whereas compound **11** have shown an IC_{50} value of 13.2 μM, 2.5 μM, 0.131 μM and 15.6 μM against cancer cell line A549, MCF7, MDA-MB231 and HepG2 respectively. Further findings of these derivatives showed that compound **10** cause cell cycle arrest of MCF7 cells in a dose dependent manner at G2/M phase. Docking analysis of target compound **10** and **11** showed that both the compound made strong interactions within the active site of protein kinase, the binding pattern of target compounds resembles as that of standard drug erlotinib, which is a potent EGFR inhibitor. In vivo acute toxicity of target compound showed that both compounds **10** and **11** are nontoxic and safe with oral LD_{50} value >500 < 2000 mg/kg which is recommended by OECD guidelines [59].

Srour et al. have reported a novel series of thiazole benzimidazole derivatives as potent inhibitor of EGFR tyrosine kinase. Target compound **12** and **13** displayed significant activity against EGFR kinase with an IC_{50} value of 71.67 nM and 109.71 nM. Both target compounds are evaluated for cytotoxicity against MCF7 cancer cell lines, compound 4n displayed an IC_{50} value of 11.91 μM and compound 4a exhibited excellent inhibitory potency with an IC_{50} value of 6.30 μM against MCF7 cancer cell line respectively. Furthermore, both compound **12** and **13** have shown good inhibition when tested against normal hTERT-RPE1 normal cells with 65 and 11.9% inhibition. Due to balanced bioactivity of target compound **13**, it is further studied for cell cycle analysis against MCF7 cell line, it displayed the cell cycle arrest at G2/M phase. Compound **13** also displayed increase in the expression of p53, Bax/Bcl-2 and caspase-3 expression and remarkable decrease in levels of PARP-1 enzyme. Molecular docking analysis of compound **12** and **13** showed that both the compounds embedded tightly by hydrogen bond formed between the Nitrogen of benzimidazole with amino acid residue Lys721 and Phe699 respectively [60].

Akhter et al. have reported as series of pyrazole benzimidazole derivatives as potential inhibitors of EGFR. Target compound **14** and **15** displayed potent activity against EGFR kinase with IC_{50} value of 0.97 μM and 1.7 μM respectively. In vitro cytotoxicity of both compound showed excellent inhibitory activity against selected cell line, compound **14** displayed an IC_{50} value of 0.97 μM, 2.2 μM and 11.9 μM and compound 5d displayed an IC_{50} value of 1.7 μM, 2.8 μM and 15.2 μM against MCF7, A549 and MDA-MB-231 cancer cell lines respectively. Target compound **14** also shown cell cycle arrest at G2/M phase of MCF7 cells by inducing apoptosis. Docking analysis of **14** displayed ability of the respective compound to fit into the active site of EGFR by forming strong hydrogen and hydrophobic within the domain (**Figure 3**) [61].

3.2.2 VEGFR 2 inhibitors

Abdullaziz et al. have reported a novel series of 2-furybenzimidazole derivatives as potent inhibitors of VEGFR-2 kinase. Target compound **16** and **17** displayed excellent

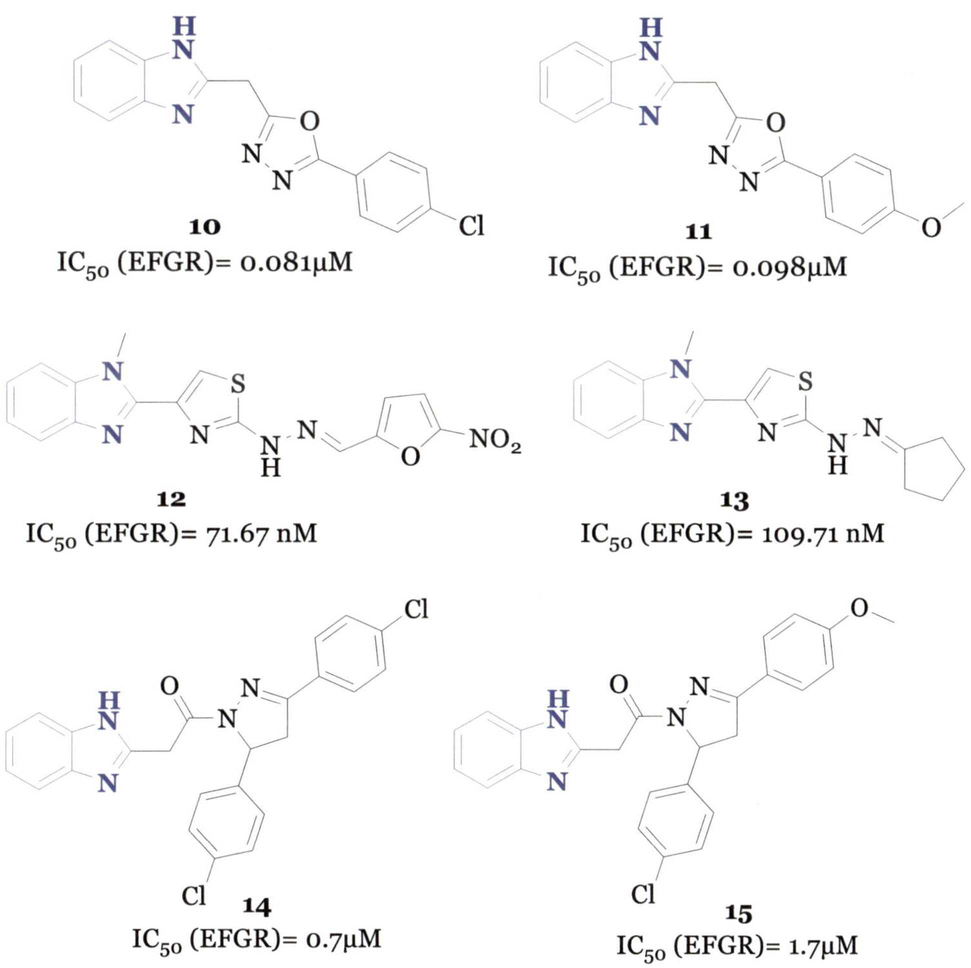

Figure 3.
Examples of benzimidazole derivatives as potent EGFR inhibitors.

inhibitory activity with total percentage inhibition of 94% and 96% and IC_{50} value of 0.64 μM and 1.26 μM compared to standard drug Sorafenib (IC_{50} value 0.1 μM) against VEGFR-2 respectively. In vitro cytotoxicity study of compound **16** and **17** displayed potential inhibitory activity with IC_{50} range of 8.33–9.86 μM against HepG2 and MCF7 cancer cell lines respectively. Molecular docking analysis of target compound showed a strong binding interaction of 2-furylbenzimidazole moiety within the active site of VEGFR-2 by involving hydrogen bond formation with key amino acid residue Glu885 and Asp1046 [62].

Lien et al. have reported novel 2-aminobenzimidazole derivative **18** as potential inhibitor of VEGFR-2. Target compound **18** exhibited 30% inhibition of kinase activity of VEGFR-2 when treated at a concentration of 10 μM. **18** displayed inhibitions of VEGF-A angiogenic action along with it also suppress MDA-MB-231 cell lines when studied in vivo. Compound **18** displayed anti-angiogenic properties by targeting VEGFR-2 signaling. Target compound **18** also found to reduce lung metastasis of B16F10 melanoma cells in mice models. Molecular docking studies of target compound showed strong binding with in the active site of VEGFR-2 by forming hydrogen bond between nitrogen of benzimidazole with amino acid residue His1026 [63].

Recently Yuan et al. have designed and synthesized a new series of benzimidazole derivatives as potent and selective inhibitor of VEGFR-2 kinase. Target compound **19** displayed excellent inhibitory activity against with VEGFR-2 kinase with an IC_{50} value of 0.054 μM, it also displayed significant anti-angiogenesis activity. In vitro cytotoxicity study of compound **19** against HepG2 and A549 cancer cell line were found promising with an IC_{50} value of 2.57 μM and 73.81 μM respectively. Cell cycle analysis of target compound **19** shows that it arrests the HepG2 cells in G0/G1 phase in a dose dependent pattern. Molecular docking analysis of compound **19** demonstrated strong interactions within the ATP binding active site of VEGFR-2 kinase [64] (**Figure 4**).

3.2.3 EGFR/VEGFR-2 dual inhibitors

Meguid et al. have reported a novel series of benzimidazole derivatives as potent dual inhibitors of EGFR and VEGFR-2 kinases. Target compound **20** and **21** displayed strong inhibitory activity against EGFR kinases, however activity against VEGFR-2 is

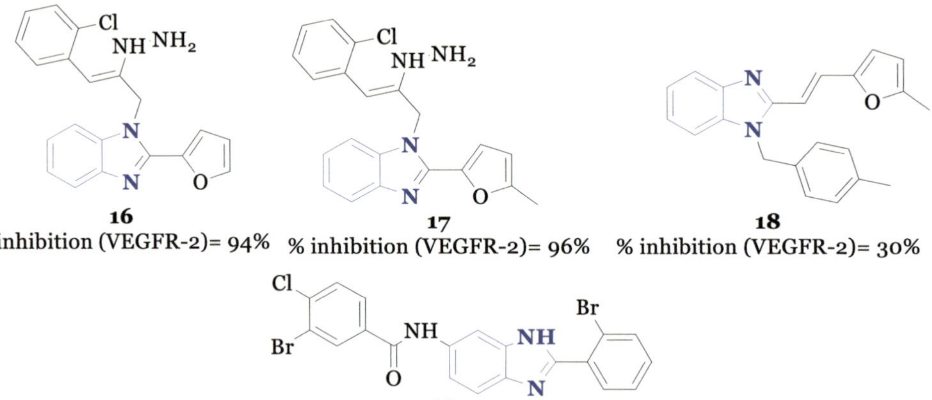

16 % inhibition (VEGFR-2)= 94%

17 % inhibition (VEGFR-2)= 96%

18 % inhibition (VEGFR-2)= 30%

19 IC_{50} (EGFR)= 0.054 μM

Figure 4.
Examples of benzimidazole derivatives as potent VEGFR-2 inhibitors.

considerably good. Target compound **20** exhibited an IC_{50} value of 0.157 μM against EGFR and 123.27 μM against VEGFR-2 kinase. Target compound **21** displayed an IC_{50} value of 0.109 μM and 69.62 μM against EGFR and VEGFR-2 kinases respectively. Cytotoxicity activity of both compound **9** and **21** was also found excellent against HeLa cancer cell line with IC_{50} value of 1.62 μM and 1.44 μM compare to standard drug doxorubicin which displayed an IC_{50} value of 2.05 μM respectively. Cell cycle analysis study revealed that both compounds arrest cell cycle of HeLa cells at G0/G1 phase. Furthermore, docking analysis showed that target compound **20** and **21** demonstrated strong binding within the active site of HER2 kinase with dock score of −9.4 and −9.7 kcal/mol respectively [65].

Kassab et al. have reported novel quinazoline bearing benzimidazole derivatives as potential inhibitors of EGFR and VEGFR-2 kinases. Target compound 22 displayed excellent inhibitory activity against EGFR kinase with an IC50 value of 127.4 μM, whereas it displayed an IC50 value of 185.7 μM against VEGFR-2 kinase. Further, cytotoxicity study of compound against MCF7 cancer cell line demonstrated good potency with IC50 value of 12.0 μM [66] (**Figure 5**).

3.2.4 PI3K inhibitors

GSK2636771 (**23**) is a novel, potent, orally available benzimidazole derivatives. It demonstrated selective PI3K beta inhibitor with antineoplastic activity. Preclinical study of GSK2636771 demonstrated selective inhibition of PTEN-deficient cancer cell growth along with inhibition of protein kinase B in a dose and time dependent manner. First in human trial study of GSK2636771 in patients of advanced solid tumors on oral administration as monotherapy demonstrated significant exposure, inhibition of target and excellent safety profile [67, 68].

Jin et al. have reported novel benzimidazole derivatives as potent PI3K inhibitor. Target compound 24 was found most potent against PI3Kα with 36% and 86% inhibition compare to reference drug Alpelisib, which showed an inhibition of 110% and 109% at 50 nM and 500 nM respectively. Further, molecular docking analysis of target compound 24 demonstrated strong binding with six strong hydrogen bond with GLN-859, SER-854 and VAL-851 amino acid residues. Further, HUMO-LUMO calculation which is studied by using Gaussian 09 software target compound 24 showed presence of thiazole core and amide bonds which played an important role in its biological activity [69].

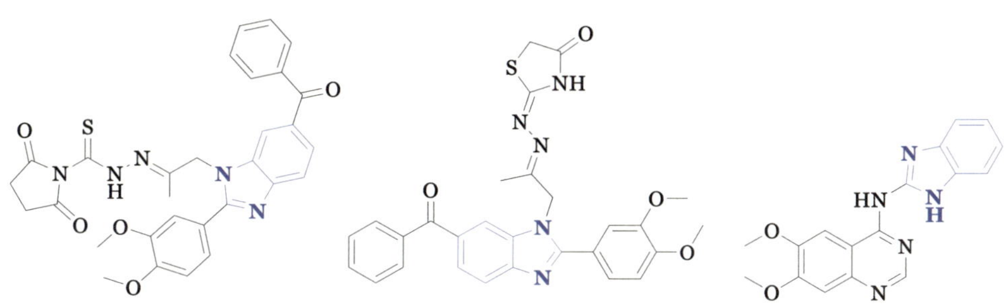

20
IC_{50} (EGFR)= 0.157 mM
IC_{50} (VEGFR-2)= 123.27 mM

21
IC_{50} (EGFR)= 0.109 mM
IC_{50} (VEGFR-2)= 69.62 mM

22
IC_{50} (EGFR)= 127.4 mM
IC_{50} (VEGFR-2)= 185.7 mM

Figure 5.
Examples of benzimidazole derivatives as potent EGFR/VEGFR dual inhibitors.

Recently a novel series of benzimidazole based dehydroabietic acid derivatives were reported Yang et al. as potent PI3Kα inhibitors. Target compound **25** have demonstrated excellent PI3K inhibitory activity with an IC_{50} value of 0.012 μM against PI3Kα which is 17-fold greater compare to PI3Kβ (IC_{50} value 0.21 μM) isoenzyme. Compound **25** is a selective PI3Kα inhibitor, it also displayed suppression of phosphorylated Akt level in HCT-116 cancer cells in a dose dependent pattern. In vitro cytotoxic activity of compound **25** showed its potent inhibitory activity against selected cancer cell line namely HCT-116, MCF-7, HeLa, HepG2 and Ges-1 cancer cell lines with an IC_{50} value of 0.18 μM, 0.43 μM, 0.71 μM, 0.63 μM and 21.95 μM respectively. Further cell apoptosis study of target compound **25** showed that it induces also apoptosis in HCT-116 when treated in a concentration dependent manner, Compound **25** comes out as potent PI3Kα inhibitor, it can be a promising agent for further development in discovery of novel anticancer agent [70].

Chanrasekhar et al. have reported a novel series of benzimidazole derivatives as potent PI3K inhibitors Target compound **26** was found to exhibit potential inhibitory activity against PI3Kβ inhibitor, it demonstrated excellent inhibitory potency with an IC_{50} value of 0.002 μM against PI3Kβ with good selectivity against all three isoforms of class I PI3Ks. Further pharmacokinetic profile of compound was evaluated in four different preclinical species (Sprague-Dawley rat, Beagle dog, Cynomolgus monkey, Rhesus monkey). Target compound **26** has shown low to intermediate clearance compare to hepatic flow of blood, whereas in rat model consistent high oral availability and high permeability was observed [71].

Wu et al. have reported a novel series of triazine substituted benzimidazole derivatives a potent dual inhibitor of PI3K and mTOR, most of the compounds from the series displayed potent inhibitory activity with IC_{50} below 33 nM. Target compound **27** was found most potent in the series, it exhibited strong inhibitory activity against both kinases with an IC_{50} value of 5.1 μM and 5.6 μM against PI3Kδ and mTOR, it exhibited PI3Kα and PI3Kβ at an IC_{50} of 7.3 nM and 21.3 nM respectively. Further, western blot analysis of compound **27** shown inhibition of phosphorylation of Akt and p70S6K, confirming dual inhibitory activity of the presenting compound. Target compound **27** displayed potent antiproliferative activity against selected cell lines, exhibited an IC50 of 0.4 μM, 0.9 μM, 1.5 μM, 7.3 μM and 7.7 μM against MCF-7, HCT116, MDA-MB-231, CNE2 and HeLa respectively. Compound **27** displayed promising PI3K/mTOR dual inhibitory activity, further development can add a potent dual inhibitor in the regimen of cancer therapy [72].

Shin et al. have reported a novel series of benzimidazole derivatives a potent inhibitor of PI3Kδ. Target compound **28** and **29** displayed an IC_{50} value of 0.016 μM and 0.019 μM against PI3Kδ and IC_{50} value of 1.78 μM and 2.33 μM PI3Kβ respectively. In vivo pharmacokinetic profile of target compound was found good with oral bioavailability of 45% and 41% respectively. In vivo studied of compound **28** and **29** suggested that both the compounds can inhibit KLH-specific antibodies [73].

He et al. has reported benzimidazole-isoquinolinone derivatives which inhibits the cell growth via inhibiting PI3K/mTOR/Akt pathway. Target compound **30** demonstrated excellent inhibitory activity against SW620 and HT29 cancer cell line with an GI_{50} value of 23.78 μM and 24.13 μM. Target compound **30** also decreases the levels of phosphorylated Akt and mTOR levels. Compound **30** also demonstrated cell cycle arrest of human colorectal cancer cells at G2/M phase by decreasing the levels of cyclin B1 and CDK1 [74].

Wu et al. have reported triazine bearing benzimidazole derivatives a potent inhibitor of PI3K and mTOR. Target compound **31** and **32** displayed potent activity with and IC_{50} value of 2.3 nM and 13.0 nM against PI3Kδ, IC_{50} value of 14.6 and 20.1 nM against PI3Kα and IC_{50} value of 34.0 and 28.0 against PI3Kβ isoform respectively. Both the compound also displayed excellent inhibitory potency against mTOR with an IC_{50}

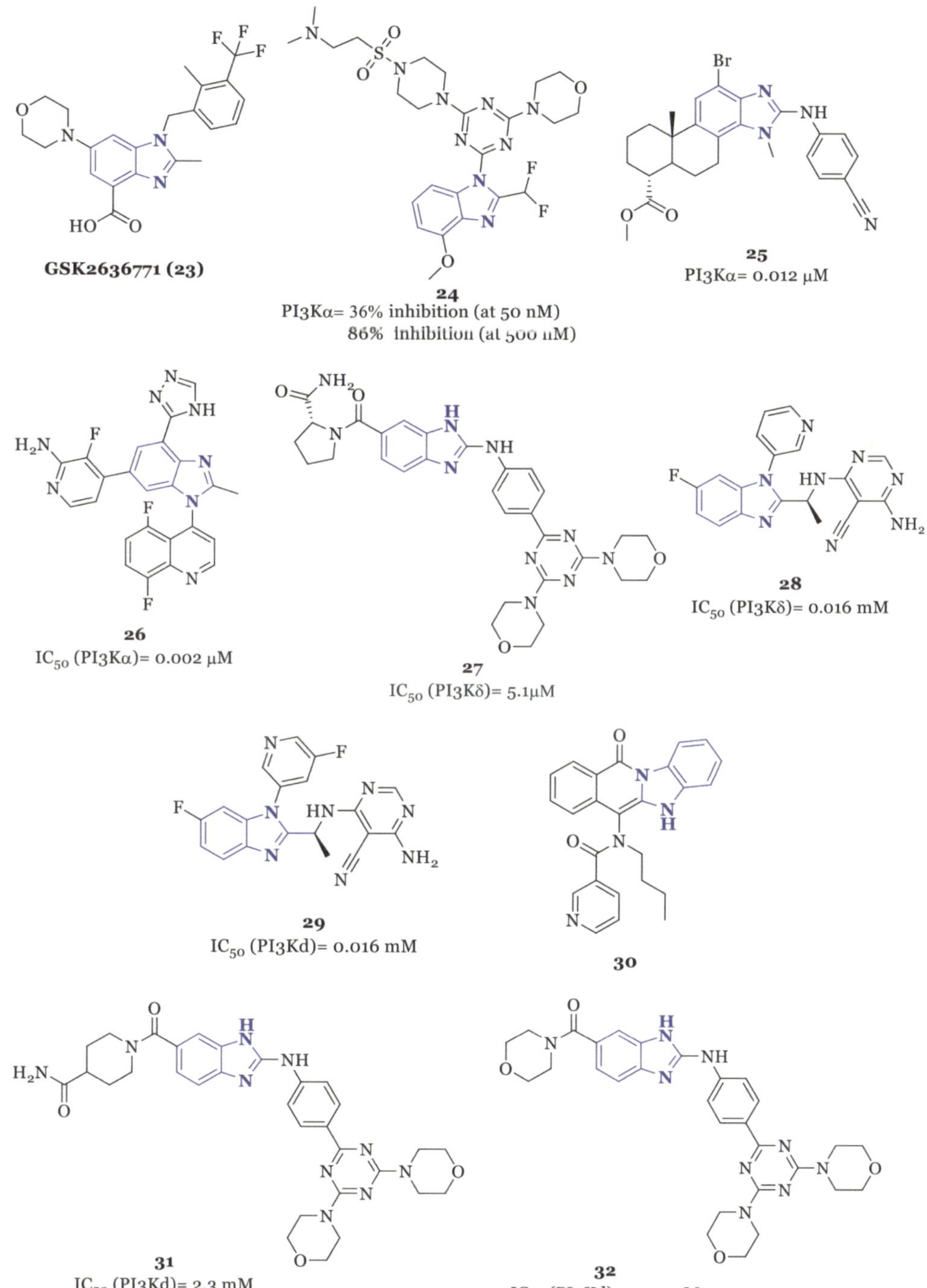

Figure 6.
Examples of benzimidazole derivatives as potent PI_3K inhibitors.

value of 12.9 nM and 15.4 nM respectively. Further, compound **32** was evaluated for antiproliferative activity where it demonstrated moderate activities against selected cancer cell line HCT116, HepG2, HeLa, MDA-MB-231 and MCF7 with an IC_{50} value of 0.3 μM, 1.3 μM, 2.4 μM, 4.8 μM and 4.9 μM respectively. Further western blot analysis study of compound **32** confirmed that it completely prevented the phosphorylation of Akt and p70S6K in HCT116 cells, thus target compound was determined as potential dual inhibitor of PI3K and mTOR kinase. Molecular docking analysis of compound **32** displayed that good binding interaction within the active site of PI3Kα [75] (**Figure 6**).

3.2.5 CDK inhibitors

Ibrahim et al. have reported a novel series of flavopiridol-benzimidazole as potent inhibitor potent inhibitor of CDK2 and CDK9 kinase. Target compound **33** exhibits potential inhibitory activity with an IC_{50} value of 0.064 and 1.725 μM against CDK2 and CDK9 kinases respectively. Furthermore, compound **33** also displayed potential antiproliferative activity against selected cancer cell line SKOV3, PC3 and K562 with an IC_{50} value of 94.0 μM, 85.0 μM and 50.8 μM respectively. Cell cycle analysis study of target compound revealed that it arrests the cell cycle of K562 cancer cell at G1 and G2 phase in a dose dependent manner [76] (**Figure 7**).

3.3 Benzimidazole based hybrid derivatives as potent anticancer agents

Pankaj et al. have reported a novel hybrid derivatives of benzimidazole-thiazolidinedione as potent cytotoxic agents. Target compound **34** demonstrated potent inhibitory activity against A549, DU-145, MDA-MB-231 and PC-3 cancer cell line with an IC_{50} value of 11.46 μM, 31.41 μM, 29.18 μM and 39.87 μM respectively. Compound **34** have shown cell cycle arrest in G2/M phase of A549 cells in a dose dependent manner. Furthermore, compound **34** also demonstrated cell shrinkage of A549 cells along with chromatin condensation and horse shoe shaped nuclei formation [77].

Sivaramakarthikeyan et al. have reported novel hybrid derivatives of benzimidazole and pyrazole as potent anticancer agents. Compound 35 and 36 have demonstrated potent anticancer activity against selected human pancreatic cancer cell lines namely SW1990 and AsPC1 with an IC50 value in range of 30.9–61.8 μM respectively. Molecular docking study of both compound showed significant binding with the active site of B-cell lymphoma [78].

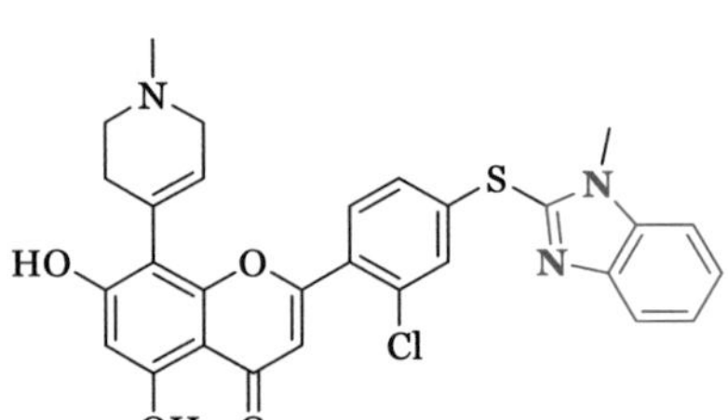

33

IC_{50} (CDK2)= 0.064□□

IC_{50}(CDK9)= 1.725 □M

Figure 7.
Examples of benzimidazole derivative as potent CDK inhibitor.

Mantu et al. have reported a novel series of benzimidazole-quinoline hybrid derivatives as potent anticancer agent. Target compound 37 exhibited potent antitumor activity against renal cancer cell line A498 and breast cancer cell line MDA-MB-468 with percentage growth inhibition of 52.92% and 56.54% respectively. Compound 37 also exhibited potent antitumor activity against leukemia cell line RPMI-8226 and non-small cell lung cancer cell line NCI-H23 with total growth inhibition of 35% [79].

Sharma et al. have reported benzimidazole-thiazolidinedione hybrid derivatives as potent anticancer agents. Target compound 38 and 39 displayed potent anticancer activity against cancer cell line with an IC50 value of in range of 0.13-10.24 μM against prostate cancer cell line PC-3, breast cancer (MDAMB-231), cervical cancer (HeLa), lung cancer (A549), and bone cancer (HT1080) cell lines. Both hybrid derivative 38 and 39 demonstrated significant inhibition of A549 cells migration

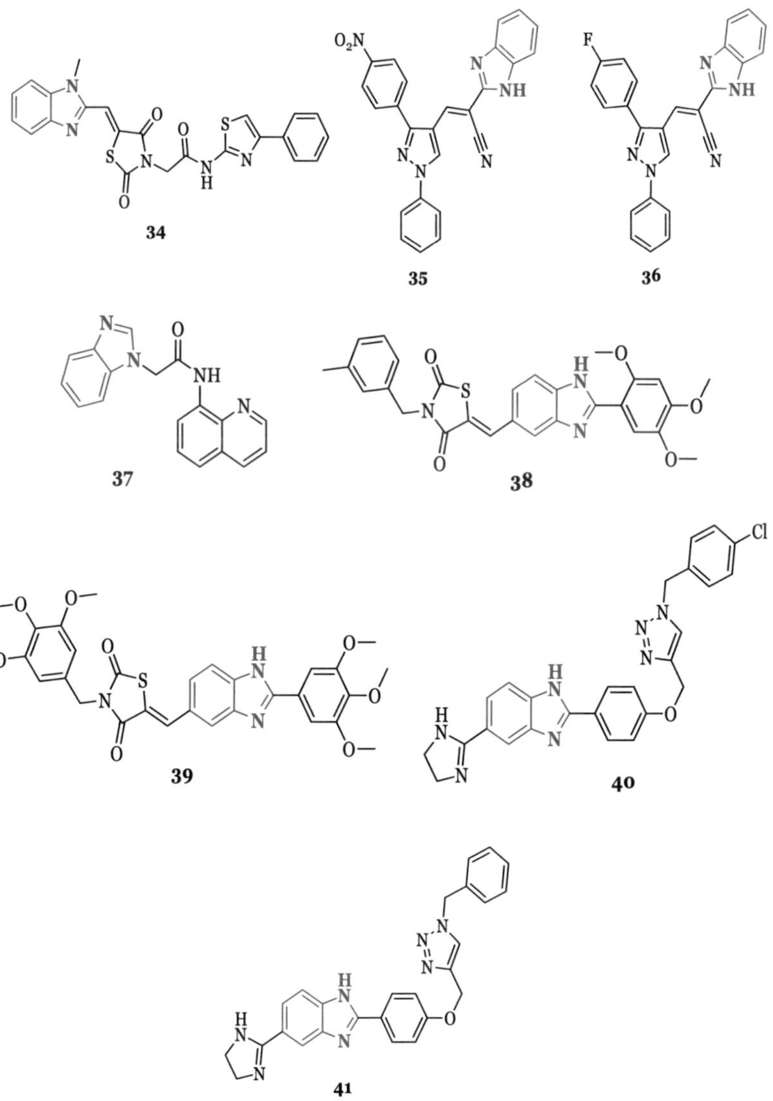

Figure 8.
Examples of benzimidazole containing hybrid derivatives as potent anticancer agents.

through disruption of F-actin assembly, further treatment with 38 and 39 also showed increase in level of ROS in A549 cells by collapsing the mitochondrial membrane potential [80].

Bistrovic et al. have reported novel hybrid derivatives of benzimidazole-1,2,3-triazole as potent anticancer agents. Target compound 40 and 41 demonstrated excellent inhibitory activity with IC50 value of 0.05 and 6.18 against A549 cancer cell line and an IC50 value of 17.53 and 8.80 against HeLa cancer cell line respectively. Furthermore, apoptosis detection study by annexin assay of compound **40** showed significant reduction of viable cell population by 70.59%, with increase in early necrotic cell population by 27.81% and late apoptotic cells by 40%. Similarly compound 41 also displayed markable decrease in cell population by 49.77%. Molecular docking analysis of compound 40 and 41 demonstrated that both the compound bind to the active site of p38 complex strongly [81] (**Figure 8**).

4. Conclusion

Many benzimidazole-containing compounds as anticancer agents are studied and available, involving various mechanisms in inhibiting mutated cancerous cells, in which kinases inhibitors play a significant role. However, in targeted therapy, benzimidazole-based derivatives are still widely explored. Due to the challenge of target specificity and poor selectivity, very few compounds have been approved to treat mutated cancers. The search for a novel benzimidazole-based next generation kinase inhibitor is going to subside such challenges. Benzimidazole-based target therapies such as enzyme inhibitors have gained a lot of attraction; owing to this, recently US FDA has approved EGFR inhibitor Abemaciclib and MEK inhibitor Binimetinib and Selumetinib as potent anticancer compounds against mutated forms of cancer. Apart from this, many benzimidazole-containing compounds are in the developmental phase as EGFR, VEGFR-2, CDK and PI3K inhibitors. However, some of the compounds demonstrated excellent kinase inhibitory activity but failed to provide a strong safety profile; these compounds will pave a path as lead compounds; further modifications, designing, and developing such compounds will give potent compounds with maximum efficiency and minimal side effects. The presented chapter mainly focuses on benzimidazole-based kinase inhibitors and their advances; the pivotal information catered here can be regarded as noteworthy and crucial by medicinal chemists for drug design, discovery and development of novel, potent and safe, target-based anticancer agents.

Acknowledgements

Authors wishes to acknowledge Jamia Hamdard (deemed-to-be-University) for providing support for conducting this study.

Conflict of interest

Authors declare "no conflict of interest."

Author details

Kashif Haider and Mohammad Shahar Yar*
Department of Pharmaceutical Chemistry, School of Pharmaceutical Education and Research (SPER), Jamia Hamdard, New Delhi, India

*Address all correspondence to: msyar@jamiahamdard.ac.in

References

[1] Siegel RL, Miller KD, Fuchs HE, Jemal A. Cancer statistics, 2021. CA: A Cancer Journal for Clinicians. 2021; **71**(1):7-33

[2] Haider K, Rahaman S, Yar MS, Kamal A. Tubulin inhibitors as novel anticancer agents: An overview on patents (2013-2018). Expert Opinion on Therapeutic Patents. 2019;**29**(8): 623-641

[3] WHO report on cancer: Setting priorities, investing wisely and providing care for all. Geneva: World Health Organization; 2020. License: CC BY-NC-SA 3.0 IGO

[4] Haider K, Rehman S, Pathak A, Najmi AK, Yar MS. Advances in 2-substituted benzothiazole scaffold-based chemotherapeutic agents. Archiv der Pharmazie. 2021;**354**:e2100246

[5] Kharb R, Haider K, Neha K, Yar MS. Aromatase inhibitors: Role in postmenopausal breast cancer. Archiv der Pharmazie. 2020;**353**(8):2000081

[6] Paul A, Singh P, Kuznetsov ML, Karmakar A, da Silva MF, Koch B, et al. Influence of anchoring moieties on new benzimidazole-based Schiff base copper (ii) complexes towards estrogen dependent breast cancer cells. Dalton Transactions. 2021;**50**(10):3701-3716

[7] Pathak A, Pandey V, Pokharel YR, Devaraji V, Ali A, Haider K, et al. Pharmacophore based drug design and synthesis of oxindole bearing hybrid as anticancer agents. Bioorganic Chemistry. 2021;**116**:105358

[8] Dandawate P, Ahmed K, Padhye S, Ahmad A, Biersack B. Anticancer active heterocyclic chalcones: Recent developments. Anti-Cancer Agents in Medicinal Chemistry (Formerly Current Medicinal Chemistry-Anti-Cancer Agents). 2021;**21**(5):558-566

[9] Scattolin T, Piccin A, Mauceri M, Rizzolio F, Demitri N, Canzonieri V, et al. Synthesis, characterization and anticancer activity of palladium allyl complexes bearing benzimidazole-based N-heterocyclic carbene (NHC) ligands. Polyhedron. 2021;**207**:115381

[10] Gondru R, Li Y, Banothu J. Coumarin–benzimidazole hybrids: A review of developments in medicinal chemistry. European Journal of Medicinal Chemistry. 2021;**227**:113921

[11] Shrivastava N, Naim MJ, Alam MJ, Nawaz F, Ahmed S, Alam O. Benzimidazole scaffold as anticancer agent: Synthetic approaches and structure–activity relationship. Archiv der Pharmazie. 2017;**350**(6):e201700040

[12] Shaharyar M, Mazumder A. Benzimidazoles: A biologically active compounds. Arabian Journal of Chemistry. 2017;**10**:S157-S173

[13] Satija G, Sharma B, Madan A, Iqubal A, Shaquiquzzaman M, Akhter M, et al. Benzimidazole based derivatives as anticancer agents: Structure activity relationship analysis for various targets. Journal of Heterocyclic Chemistry. 2021

[14] Akhtar W, Khan MF, Verma G, Shaquiquzzaman M, Rizvi MA, Mehdi SH, et al. Therapeutic evolution of benzimidazole derivatives in the last quinquennial period. European Journal of Medicinal Chemistry. 2017;**126**:705-753

[15] Akhtar MJ, Yar MS, Sharma VK, Khan AA, Ali Z, Haider MD, et al. Recent progress of benzimidazole hybrids for

anticancer potential. Current Medicinal Chemistry. 2020;**27**(35):5970-6014

[16] Gaba M, Mohan C. Development of drugs based on imidazole and benzimidazole bioactive heterocycles: Recent advances and future directions. Medicinal Chemistry Research. 2016;**25**(2):173-210

[17] Song D, Ma S. Recent development of benzimidazole-containing antibacterial agents. ChemMedChem. 2016;**11**(7): 646-659

[18] Farahat AA, Ismail MA, Kumar A, Wenzler T, Brun R, Paul A, et al. Indole and benzimidazole bichalcophenes: Synthesis, DNA binding and antiparasitic activity. European Journal of Medicinal Chemistry. 2018;**143**:1590-1596

[19] Morcoss MM, El Shimaa MN, Ibrahem RA, Abdel-Rahman HM, Abdel-Aziz M, Abou El-Ella DA. Design, synthesis, mechanistic studies and in silico ADME predictions of benzimidazole derivatives as novel antifungal agents. Bioorganic Chemistry. 2020;**101**: 103956

[20] Veerasamy R, Roy A, Karunakaran R, Rajak H. Structure–activity relationship analysis of benzimidazoles as emerging anti-inflammatory agents: An overview. Pharmaceuticals. 2021;**14**(7):663

[21] Eswayah A, Khaliel S, Saad S, Shebani N, Fhid O, Belaid A, et al. Synthesis and analgesic activity evaluation of some new benzimidazole derivatives. American Journal of Chemistry and Application. 2017;**4**(5): 30-35

[22] Kanwal A, Ahmad M, Aslam S, Naqvi SA, Saif MJ. Recent advances in antiviral benzimidazole derivatives: A mini review. Pharmaceutical Chemistry Journal. 2019;**53**(3):179-187

[23] Araujo DM, Maste MM, Alegaon S, Saxena A. Synthesis, antitubercular evaluation and docking studies of novel benzimidazole analogues. International Journal of Pharmaceutical Sciences and Research. 2018;**9**:3696-3704

[24] Matsubara Y, Matsumoto T, Yoshiya K, Yoshida A, Ikeda S, Furuyama T, et al. Budding uninhibited by benzimidazole-1 insufficiency prevents acute renal failure in severe sepsis by maintaining anticoagulant functions of vascular endothelial cells. Shock. 2019;**51**(3):364-371

[25] Wang XJ, Xi MY, Fu JH, Zhang FR, Cheng GF, You QD. Synthesis, biological evaluation and SAR studies of benzimidazole derivatives as H1-antihistamine agents. Chinese Chemical Letters. 2012;**23**(6):707-710

[26] Aroua LM, Almuhaylan HR, Alminderej FM, Messaoudi S, Chigurupati S, Al-Mahmoud S, et al. A facile approach synthesis of benzoylaryl benzimidazole as potential α-amylase and α-glucosidase inhibitor with antioxidant activity. Bioorganic Chemistry. 2021;**114**:105073

[27] Ganie AM, Dar AM, Khan FA, Dar BA. Benzimidazole derivatives as potential antimicrobial and antiulcer agents: A mini review. Mini Reviews in Medicinal Chemistry. 2019;**19**(16): 1292-1297

[28] Djemoui A, Naouri A, Ouahrani MR, Djemoui D, Lahcene S, Lahrech MB, et al. A step-by-step synthesis of triazole-benzimidazole-chalcone hybrids: Anticancer activity in human cells+. Journal of Molecular Structure. 2020; **1204**:127487

[29] Ren Y, Wang Y, Li G, Zhang Z, Ma L, Cheng B, et al. Discovery of novel benzimidazole and indazole analogues as

tubulin polymerization inhibitors with potent anticancer activities. Journal of Medicinal Chemistry. 2021;**64**(8): 4498-4515

[30] Akhtar S, Abbas M, Naeem K, Faheem M, Nadeem H, Mehmood A. Benzimidazole derivative ameliorates opioid-mediated tolerance during anticancer-induced neuropathic pain in mice. Anti-Cancer Agents in Medicinal Chemistry (Formerly Current Medicinal Chemistry-Anti-Cancer Agents). 2021;**21**(3):365-371

[31] Choi HS, Ko YS, Jin H, Kang KM, Ha IB, Jeong H, et al. Anticancer effect of benzimidazole derivatives, especially mebendazole, on triple-negative breast cancer (TNBC) and radiotherapy-resistant TNBC in vivo and in vitro. Molecules. 2021;**26**(17):5118

[32] Shirley M. Encorafenib and binimetinib: First global approvals. Drugs. 2018;**78**(12):1277-1284

[33] Kopetz S, Grothey A, Yaeger R, Van Cutsem E, Desai J, Yoshino T, et al. Encorafenib, binimetinib, and cetuximab in BRAF V600E–mutated colorectal cancer. New England Journal of Medicine. 2019;**381**(17):1632-1643

[34] Grothey A, Tabernero J, Taieb J, Yaeger R, Yoshino T, Maiello E, et al. LBA-5 ANCHOR CRC: A single-arm, phase 2 study of encorafenib, binimetinib plus cetuximab in previously untreated BRAF V600E-mutant metastatic colorectal cancer. Annals of Oncology. 2020;**31**:S242-S243

[35] Balfour JA, Goa KL. Bendamustine. Drugs. 2001;**61**(5):631-638

[36] Yamshon S, Martin P. Are novel agents ready to assume the mantle in the frontline treatment of mantle cell lymphoma? Clinical Advances in Hematology & Oncology: H&O. 2021;**19**(6):376-382

[37] Markham A, Keam SJ. Selumetinib: First approval. Drugs. 2020;**80**:931-937

[38] Casey D, Demko S, Sinha A, Mishra-Kalyani PS, Shen YL, Khasar S, et al. FDA Approval Summary: Selumetinib for Plexiform Neurofibroma. Clinical Cancer Research. 2021

[39] Gross AM, Wolters PL, Dombi E, Baldwin A, Whitcomb P, Fisher MJ, et al. Selumetinib in children with inoperable plexiform neurofibromas. New England Journal of Medicine. 2020;**382**(15):1430-1442

[40] Mukhopadhyay S, Maitra A, Choudhury S. Selumetinib: The first ever approved drug for neurofibromatosis-1 related inoperable plexiform neurofibroma. Current Medical Research and Opinion. 2021;**37**(5):789-794

[41] Kim ES. Abemaciclib: First global approval. Drugs. 2017;**77**(18):2063-2070

[42] Cuyun Carter G, Sheffield KM, Gossai A, Huang YJ, Zhu YE, Bowman L, et al. Real-world treatment patterns and outcomes of abemaciclib for the treatment of HR+, HER2-metastatic breast cancer. Current Medical Research and Opinion. 2021;**1**:37

[43] Toi M, Inoue K, Masuda N, Iwata H, Sohn J, Park IH, et al. Abemaciclib in combination with endocrine therapy for East Asian patients with HR+, HER2– advanced breast cancer: MONARCH 2 & 3 trials. Cancer Science. 2021;**112**(6): 2381

[44] Ghisoni E, Giannone G, Tuninetti V, Genta S, Scotto G, Aglietta M, et al. Veliparib: A new therapeutic option in ovarian cancer? Future Oncology. 2019;**15**(17):1975-1987

[45] Boussios S, Karihtala P, Moschetta M, Abson C, Karathanasi A, Zakynthinakis-Kyriakou N, et al. Veliparib in ovarian cancer: A new synthetically lethal therapeutic approach. Investigational New Drugs. 2020;**38**(1):181-193

[46] Ghisoni E, Giannone G, Tuninetti V, Genta S, Scotto G, Aglietta M, et al. Veliparib: A new therapeutic option in ovarian cancer? Future Oncology. 2019;**15**(17):1975-1987

[47] Nightingale J, Lum B, Ladwa R, Simpson F, Panizza B. Adenoid cystic carcinoma: A review of clinical features, treatment targets and advancement in improving the immune response to monoclonal antibody therapy. Biochimica et Biophysica Acta (BBA)-Reviews on Cancer. 2021;**1875**:188523

[48] Choi YJ, Kim HS, Park SH, Kim BS, Kim KH, Lee HJ, et al. Phase II study of dovitinib in patients with castration-resistant prostate cancer (KCSG-GU11-05). Cancer Research and Treatment: Official Journal of Korean Cancer Association. 2018;**50**(4):1252

[49] Gurnari C, Voso MT, Maciejewski JP, Visconte V. From bench to bedside and beyond: Therapeutic scenario in acute myeloid leukemia. Cancers. 2020;**12**(2):357

[50] Yu J, Jiang PY, Sun H, Zhang X, Jiang Z, Li Y, et al. Advances in targeted therapy for acute myeloid leukemia. Biomarker Research. 2020;**8**:1-1

[51] Chua CC, Wei AH. Future developments: Novel agents. In: Acute Myeloid Leukemia. Cham: Springer; 2021. pp. 293-315

[52] Chen J, Li N, Liu B, Ling J, Yang W, Pang X, et al. Pracinostat (SB939), a histone deacetylase inhibitor, suppresses breast cancer metastasis and growth by inactivating the IL-6/STAT3 signalling pathways. Life Sciences. 2020;**248**:117469

[53] Njar VC, Brodie AM. Discovery and development of Galeterone (TOK-001 or VN/124-1) for the treatment of all stages of prostate cancer. Journal of Medicinal Chemistry. 2015;**58**(5): 2077-2087

[54] Xu Y, Liao S, Wang L, Wang Y, Wei W, Su K, et al. Galeterone sensitizes breast cancer to chemotherapy via targeting MNK/eIF4E and β-catenin. Cancer Chemotherapy and Pharmacology. 2021;**87**(1):85-93

[55] Cui J, Xiao Z, Zhang LL. Clinical efficacy and safety of nazartinib for epidermal growth factor receptor mutated non-small cell lung cancer: Study protocol for a prospective, multicenter, open-label. Medicine. 2021;**100**(21):e25992

[56] Tan DS, Leighl NB, Riely GJ, Yang JC, Sequist LV, Wolf J, et al. Safety and efficacy of nazartinib (EGF816) in adults with EGFR-mutant non-small-cell lung carcinoma: A multicentre, open-label, phase 1 study. The Lancet Respiratory Medicine. 2020;**8**(6):561-572

[57] Jassem J, Dziadziuszko R. Nazartinib in EGFR Thr790Met-mutant non-small-cell lung cancer. The Lancet Respiratory Medicine. 2020;**8**(6):528-529

[58] Liu C, Lu H, Wang H, Loo A, Zhang X, Yang G, et al. Combinations with allosteric SHP2 inhibitor TNO155 to block receptor tyrosine kinase signaling. Clinical Cancer Research. 2021;**27**(1):342-354

[59] Akhtar MJ, Siddiqui AA, Khan AA, Ali Z, Dewangan RP, Pasha S, et al. Design, synthesis, docking and QSAR study of substituted benzimidazole linked oxadiazole as cytotoxic agents,

EGFR and erbB2 receptor inhibitors. European Journal of Medicinal Chemistry. 2017;**126**:853-869

[60] Srour AM, Ahmed NS, Abd El-Karim SS, Anwar MM, El-Hallouty SM. Design, synthesis, biological evaluation, QSAR analysis and molecular modelling of new thiazol-benzimidazoles as EGFR inhibitors. Bioorganic & Medicinal Chemistry. 2020;**28**(18):115657

[61] Akhtar MJ, Khan AA, Ali Z, Dewangan RP, Rafi M, Hassan MQ, et al. Synthesis of stable benzimidazole derivatives bearing pyrazole as anticancer and EGFR receptor inhibitors. Bioorganic Chemistry. 2018;**78**:158-169

[62] Abdullaziz MA, Abdel-Mohsen HT, El Kerdawy AM, Ragab FA, Ali MM, Abu-Bakr SM, et al. Design, synthesis, molecular docking and cytotoxic evaluation of novel 2-furybenzimidazoles as VEGFR-2 inhibitors. European Journal of Medicinal Chemistry. 2017;**136**: 315-329

[63] Lien JC, Chung CL, Huang TF, Chang TC, Chen KC, Gao GY, et al. A novel 2-aminobenzimidazole-based compound Jzu 17 exhibits anti-angiogenesis effects by targeting VEGFR-2 signalling. British Journal of Pharmacology. 2019;**176**(20):4034-4049

[64] Yuan X, Yang Q, Liu T, Li K, Liu Y, Zhu C, et al. Design, synthesis and in vitro evaluation of 6-amide-2-aryl benzoxazole/benzimidazole derivatives against tumor cells by inhibiting VEGFR-2 kinase. European Journal of Medicinal Chemistry. 2019;**179**:147-165

[65] Abd El-Meguid EA, El-Deen EM, Nael MA, Anwar MM. Novel benzimidazole derivatives as anti-cervical cancer agents of potential multi-targeting kinase inhibitory activity. Arabian Journal of Chemistry. 2020;**13**(12):9179-9195

[66] Kassab AE, Gedawy EM, El-Nassan HB. Synthesis of 4-heteroaryl–quinazoline derivatives as potential anti-breast cancer agents. Journal of Heterocyclic Chemistry. 2017;**54**(1): 624-633

[67] Mateo J, Ganji G, Burris HA, Han SW, Swales K, DeYoung P, et al. A first time in human trial of GSK2636771, a PI3Kβ selective inhibitor, in patients with advanced solid tumors

[68] Mateo J, Ganji G, Lemech C, Burris HA, Han SW, Swales K, et al. A first-time-in-human study of GSK2636771, a phosphoinositide 3 kinase beta-selective inhibitor, in patients with advanced solid tumors. Clinical Cancer Research. 2017;**23**(19):5981-5992

[69] Jin RY, Tang T, Zhou S, Long X, Guo H, Zhou J, et al. Design, synthesis, antitumor activity and theoretical calculation of novel PI3Ka inhibitors. Bioorganic Chemistry. 2020;**98**:103737

[70] Yang YQ, Chen H, Liu QS, Sun Y, Gu W. Synthesis and anticancer evaluation of novel 1H-benzo [d] imidazole derivatives of dehydroabietic acid as PI3Kα inhibitors. Bioorganic Chemistry. 2020;**100**:103845

[71] Chandrasekhar J, Dick R, Van Veldhuizen J, Koditek D, Lepist EI, McGrath ME, et al. Atropisomerism by design: Discovery of a selective and stable phosphoinositide 3-kinase (PI3K) β inhibitor. Journal of Medicinal Chemistry. 2018;**61**(15):6858-6868

[72] Wu TT, Guo QQ, Chen ZL, Wang LL, Du Y, Chen R, et al. Design, synthesis and bioevaluation of novel substituted triazines as potential dual PI3K/mTOR inhibitors. European Journal of Medicinal Chemistry. 2020;**204**:112637

[73] Shin Y, Suchomel J, Cardozo M, Duquette J, He X, Henne K, et al.

Discovery, optimization, and in vivo evaluation of benzimidazole derivatives AM-8508 and AM-9635 as potent and selective PI3Kδ inhibitors. Journal of Medicinal Chemistry. 2016;**59**(1):431-447

[74] He LJ, Yang DL, Li SQ, Zhang YJ, Tang Y, Lei J, et al. Facile construction of fused benzimidazole-isoquinolinones that induce cell-cycle arrest and apoptosis in colorectal cancer cells. Bioorganic and Medicinal Chemistry. 2018;**26**(14):3899-3908

[75] Wu TT, Guo QQ, Chen ZL, Wang LL, Du Y, Chen R, et al. Design, synthesis and bioevaluation of novel substituted triazines as potential dual PI3K/mTOR inhibitors. European Journal of Medicinal Chemistry. 2020;**204**:112637

[76] Ibrahim N, Bonnet P, Brion JD, Peyrat JF, Bignon J, Levaique H, et al. Identification of a new series of flavopiridol-like structures as kinase inhibitors with high cytotoxic potency. European Journal of Medicinal Chemistry. 2020;**199**:112355

[77] Sharma P, Reddy TS, Thummuri D, Senwar KR, Kumar NP, Naidu VG, et al. Synthesis and biological evaluation of new benzimidazole-thiazolidinedione hybrids as potential cytotoxic and apoptosis inducing agents. European Journal of Medicinal Chemistry. 2016;**124**:608-621

[78] Sivaramakarthikeyan R, Iniyaval S, Saravanan V, Lim WM, Mai CW, Ramalingan C. Molecular hybrids integrated with benzimidazole and pyrazole structural motifs: Design, synthesis, biological evaluation, and molecular docking studies. ACS Omega. 2020;**5**(17):10089-10098

[79] Mantu D, Antoci V, Moldoveanu C, Zbancioc G, Mangalagiu II. Hybrid imidazole (benzimidazole)/pyridine (quinoline) derivatives and evaluation of their anticancer and antimycobacterial activity. Journal of Enzyme Inhibition and Medicinal Chemistry. 2016;**31** (suppl. 2):96-103

[80] Sharma P, Reddy TS, Kumar NP, Senwar KR, Bhargava SK, Shankaraiah N. Conventional and microwave-assisted synthesis of new 1H-benzimidazole-thiazolidinedione derivatives: A potential anticancer scaffold. European Journal of Medicinal Chemistry. 2017;**138**:234-245

[81] Bistrović A, Krstulović L, Harej A, Grbčić P, Sedić M, Koštrun S, et al. Design, synthesis and biological evaluation of novel benzimidazole amidines as potent multi-target inhibitors for the treatment of non-small cell lung cancer. European Journal of Medicinal Chemistry. 2018;**143**:1616-1634

Chapter 2

The Anticancer Profile of Benzimidazolium Salts and Their Metal Complexes

Imran Ahmad Khan, Noor ul Amin Mohsin, Sana Aslam and Matloob Ahmad

Abstract

Cancer is the most lethal ailment throughout the world in the present era. The development of new anticancer remedies with minor unhealthful effects and an alternate mechanism is crucial. Benzimidazole is a distinguished heterocyclic compound and is now recognized as the privileged scaffold for new drug discovery. This chapter deals with the anticancer capability of benzimidazolium salts and their metal complexes. The benzimidazolium derivatives have been prepared by the introduction of aliphatic and aromatic groups at two nitrogen atoms of the benzimidazole ring. Other modifications include hybridization with other pharmacophores and the preparation of metal complexes. The potent derivatives presented in this review can serve as novel drug candidates against cancer.

Keywords: benzimidazolium salts, Benzimidazole, metal complexes, salts, hybrid, silver, breast cancer (MCF-7) cell line, colon cancer (HCT-116) cell line

1. Introduction

Cancer is among the most dreadful diseases and a significant cause of assassinations all over the globe. In 2018, 9.6 million expirations were because of this malady [1]. Breast cancer is the paramount form of cancer in women all over the world. In 2018, 2.3 million victims and 627,000 fatalities were reported due to breast cancer. Prostate cancer is the second most common cancer in males and 1.3 million patients were reported in 2018 [1]. It has been deduced that more than 13 million people will die due to cancer in 2030 [2]. The major risk factors associated with cancer are chronic infections, inherited mutations in genes, overweight, no physical activity, exposure to ionizing radiation, and carcinogens such as polychlorinated biphenyls, chloroform, Dichlorodiphenyltrichloroethane (DDT), and formaldehyde [3]. Treatment patterns for cancer involve radiotherapy, surgery, and drug therapy. Drug therapy includes inorganic, organic, organometallic monomers, and polymers as well as nanoparticles [4]. Drug therapy is associated with severe adverse properties such as alopecia, anemia, and infertility. There is also the development of resistance against currently

IntechOpen

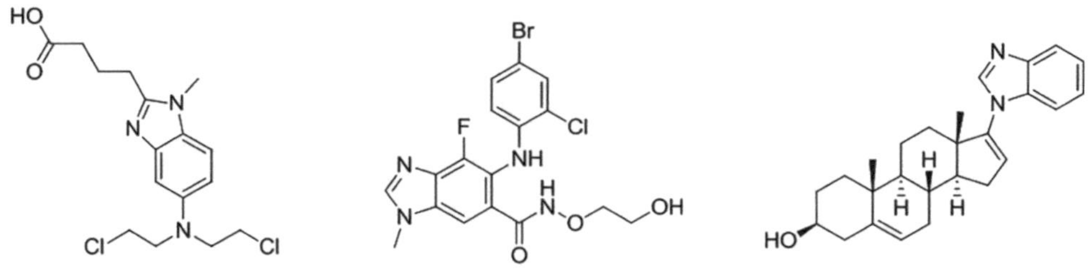

Bendamustine **(1)** Selumetinib **(2)** Galeterone **(3)**

Figure 1.
Anticancer drugs based on benzimidazole scaffold.

available drugs [5]. Consequently, the development of new anticancer drugs affiliated with low toxicities is very significant.

Nitrogen-containing heterocycles are abundantly present in natural and synthetic drug molecules [6]. Benzimidazole is one of the most significant members of nitrogen-containing heterocycles. This heterocycle is a constituent of the structures of some natural compounds such as vitamin B12 [7]. Benzimidazole derivatives have antihypertensive [8], anti-inflammatory [9], antimicrobial [10], antiulcer [11], antiviral [12], antioxidant [13], antitumor [14], lipid modulator, and anticoagulant properties [15]. Benzimidazole derivatives have also the major therapeutic activities against cancer [16–18]. Benzimidazole is also the main pharmacophore of anticancer drugs (**Figure 1**) such as bendamustine (**1**), selumetinib (**2**), and galeterone (**3**) out of which bendamustine is approved for clinical use, while the other two are in clinical trial stages [19, 20]. The structural resemblance of benzimidazole with nucleotides makes them very vital from the biological point of view [21]. Benzimidazolium salts are 1,3-disubstituted benzimidazole derivatives and possess acidic hydrogen at position 2. Benzimidazole salts find application as a carbene precursor for the preparation of n-heterocyclic carbenes (NHC) with different metals [22]. These benzimidazolium salts and their complexes have displayed significant antimicrobial and anticancer properties [23]. This review deals with the anticancer activities of benzimidazolium salts and their metal complexes.

2. Anticancer properties of benzimidazolium salts

Benzimidazolium salts and their anticancer capabilities have been reviewed in the following sections.

2.1 Hybrid molecules containing benzimidazolium salts

Molecular hybridization has become an effective approach for new drug discovery. In molecular hybridization, two or more pharmacophores are linked to each other to produce the new molecules [24, 25]. Yang et al. synthesized hybrid molecules in which benzimidazolium salts were linked to trimethoxy phenyl chalcones. Compound **4** (**Figure 2**) demonstrated excellent anticancer potential against leukemia (HL-60), breast carcinoma (MCF-7), and colon carcinoma (SW480) cell lines presenting IC_{50} values of 0.83, 1.57, and 2.92 μM, respectively, which is 5–11-folds higher than the standard drug cisplatin. In compound **4**, 2-naphthylmethyl substituent is attached

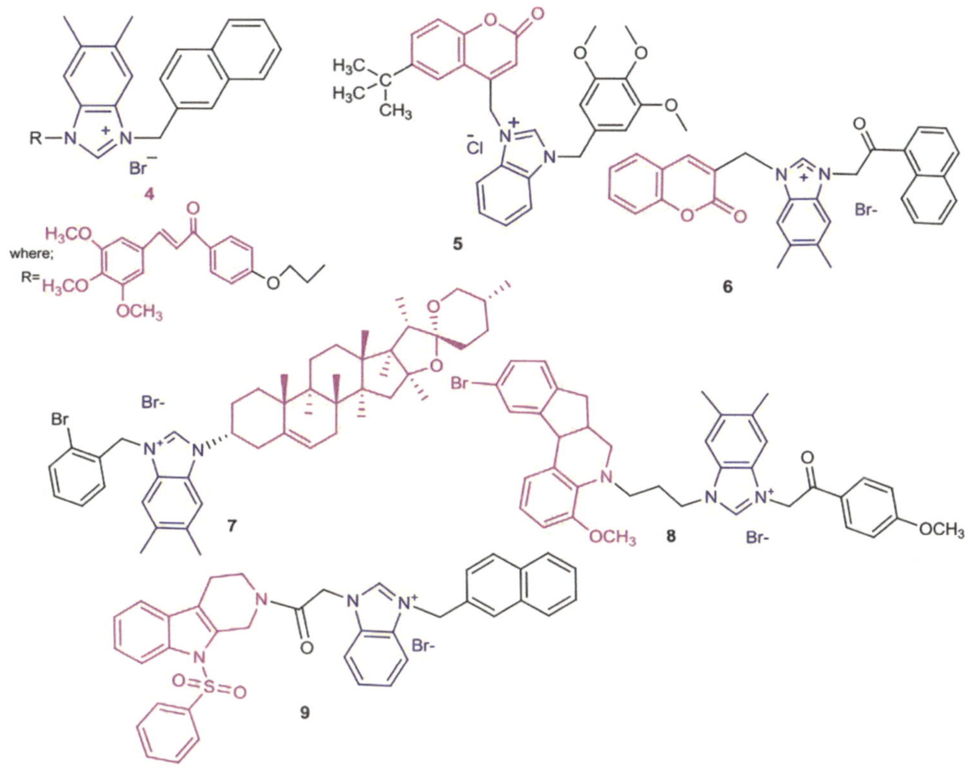

Figure 2.
Hybrid molecules comprising benzimidazole salts and natural compounds.

to benzimidazole nitrogen. The superior activity of these salts was related to the high solubility of benzimidazolium salts. Benzimidazole derivatives with substituents at positions 5, 6 showed greater activity as compared to unsubstituted benzimidazoles [26]. Karatas et al. reported a series of hybrid molecules in which coumarin was attached to the substituted benzimidazolium chlorides. Anticancer screening by 3-(4,5-dimethylthiazol-2-yl)-2,5-diphenyl tetrazolium bromide (MTT) assay showed that these compounds have the potential to discontinue the cell cycle for human prostate (PC-3) and ovarian (A2780) cancer cells. Specifically, compound **5** (IC_{50} = 44.5 μM) showed some inhibitory potential against PC-3 at the dose of 1 μM [27]. Next year, Wang et al. produced the 3-benzyl coumarin imidazolium salts using the hybrid molecular strategy. The anticancer activity evaluation showed that compound **6** is one of the most active derivatives having IC_{50} values ranging from 2.04 to 4.51 μM. The best activity was displayed *versus* the MCF-7 cell line having an IC_{50} value of 2.04 μM. The 5, 6-dimethyl-substituted benzimidazole hybrids exhibited prominent activity as compared to unsubstituted benzimidazoles or imidazoles. These molecules showed selectivity for MCF-7 and SW-480 cancer cell lines. Compound **6** also showed the apoptosis in liver carcinoma (SMMC-7721) cell line [28]. Deng et al. linked benzimidazolium salts with steroidal molecules such as cholesterol, dehydroepiandrosterone, and diosgenin. Anticancer activities were carried out against HL-60, A-549, SMMC-7721, SW480, and MCF-7 cancer cell lines. Diosgenin-imidazolium salts displayed higher activity, and compound **7** was the most effective having IC_{50} values from 0.44 to 0.79 μM against different cell lines.

Compound **7** disrupted the cell cycle in G1/G0 stage and showed apoptosis in the SMMC-7721 cell line. Structure-activity relationship (SAR) studies showed that 5, 6-dimethyl-substituted benzimidazolium salts showed excellent anticancer activity. Attachment of 2-bromobenzyl with 2-naphthyl methyl at position 3 of benzimidazole also amplified the activity [29]. Brazilin is a natural compound and possesses extensive bioactivities such as anti-inflammatory, anticancer, and antioxidant [30]. Huang et al. connected aza-brazilin with imidazolium salts to produce hybrid molecules. Anticancer activity was evaluated against A549, SMMC-7721, MCF-7, and SW480 cell lines. Derivative **8** appeared as the most active (IC_{50} = 0.35 μM) against the MCF-7 cell line and displayed more potency as compared to cisplatin. Derivatives having a 5,6-dimethylbenzimidazole ring displayed prominent activity. The introduction of electron-withdrawing groups on the aza-brazilin nucleus produced more active derivatives. Derivatives having an alkyl chain as linker groups produce higher potency as compared to acyl chains [31]. Zhou et al. also used a hybrid molecular strategy to conjugate *N*-substituted tetrahydro-*β*-carbolines with imidazolium salts. Compound **9** exhibited prominent activity with IC_{50} values ranging from 2.61 to 17.13 μM against five different cancer cell lines. Most prominent activity was achieved against MCF-7 (IC_{50} = 2.79 μM) and SW-480 (IC_{50} = 9.46 μM) cancer cell lines. Compound **9** carries a naphthyl methyl scaffold at position 3 of benzimidazole. Compound **9** also showed the phenomenon of apoptosis in the MCF-7 cell line as well as inhibited the cell cycle in the G1 phase [32].

Xu et al. coupled a three-substituted indole ring with imidazolium salts to produce new hybrid molecules. Upon evaluation of anticancer activity, compound **10** (**Figure 3**) showed prominent performance against MCF-7 (IC_{50} = 3.19 μM), A549 (IC_{50} = 3.51 μM), SW480 (IC_{50} = 11.57 μM), and SMMC-7721 (IC_{50} = 3.60 μM) cancer cell lines. SAR studies showed that 5,6-dimethyl-substituted benzimidazole derivatives showed prominent activity as compared to unsubstituted benzimidazoles. Compound **10** carries a naphthyl acyl ring at position 3 of the benzimidazole nucleus. Further studies showed that compound **10** is capable of inducing apoptosis and caused cell cycle blockage in the *S* phase [33]. Li et al. synthesized carbazole and imidazolium salts using the molecular hybridization technique. Replacement of the imidazole ring by the benzimidazole increased the anticancer selectivity against a particular cell line. Among the benzimidazolium salts, derivative **11** showed excellent activity against the HeLa cell line (IC_{50} = 0.02 μM) as compared to standard drug cisplatin (IC_{50} = 13.61 μM) [34]. Wang et al. synthesized imidazolium salts and dibenzofuran comprising hybrid molecules. The estimation of anticancer activity against five cancer cell lines showed that 2-methylbenzimidazolium and dibenzofuran hybrid molecular salts are more active as compared to individual molecules. 2-Methyl-substituted benzimidazolium salts showed higher activity as compared to unsubstituted and 5,6-dimethylbenzimidazolium salts. Compound **12** expressed prominent activity (IC_{50} = 0.64–1.47 μM) against MCF-7, A549, SW480, HL-60, and SMMC-7721 cancer cell lines. Most prominent activity was observed against MCF-7 (IC_{50} = 0.64 μM) and SW-480 (IC_{50} = 0.88 μM) cell lines. Compound **12** carries a 4-methoxy phenacyl substituent at position 3 of the benzimidazole nucleus. Replacement of 4-methoxy phenacyl substituent by 2-naphthylacyl also produced potent derivatives [35]. Zhang and co-workers synthesized hybrid molecules comprising 2,3,6,7-tetrahydrobenzodifuran and imidazolium salts. Compound **13** appeared as the most active derivative against five cancer cell lines having an IC_{50} value less than 4.34 μM. Compound **13** showed selectivity for A549, SMMC-7721, and SW-480 cancer cell lines. Some derivatives of this series also exhibited the phenomenon of apoptosis and seized

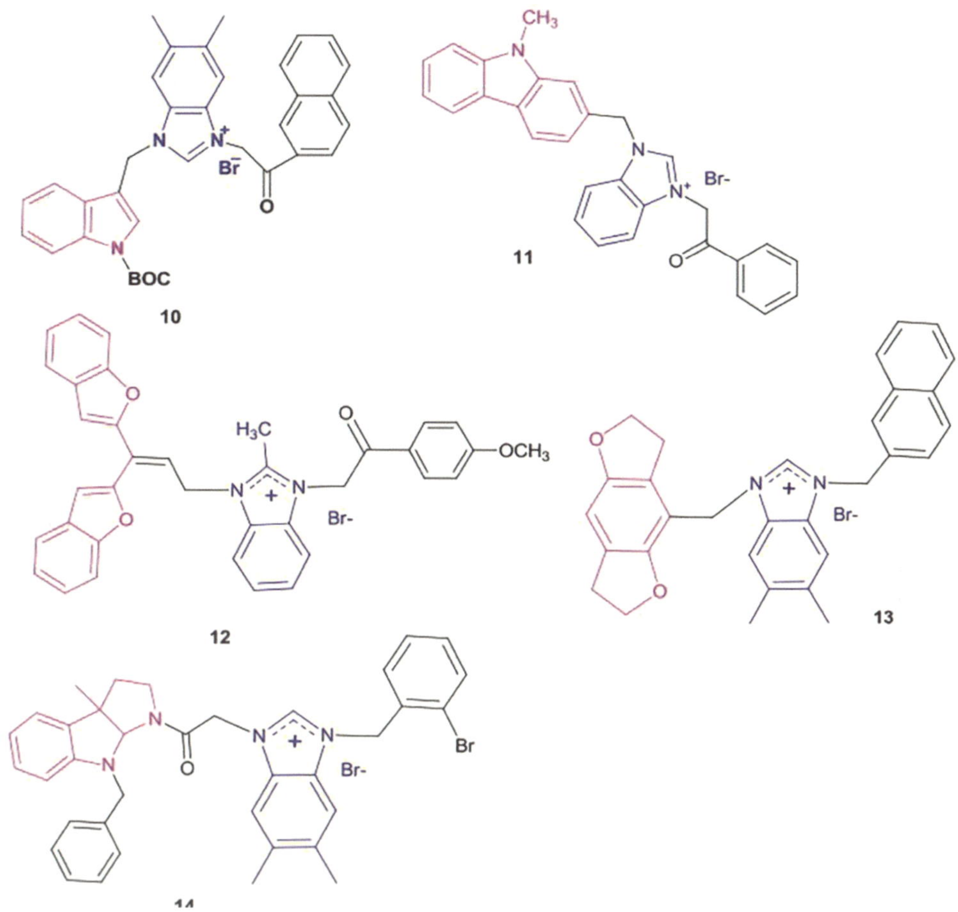

Figure 3.
Hybrid molecules of benzimidazole salts with synthetic molecules.

cell cycle in the G1 stage [36]. Zhou et al. manufactured hexahydropyrrolo[2,3b]indole-1*H*-imidazolium salts as anticancer agents. The nitrogen at position 3 of the benzimidazole ring was linked to 2-bromobenzyl and 2-naphthyl methyl scaffolds. Compound **14** emerged as the most active derivative against HL-60, MCF-7, SMMC-7721, A549, and SW-480 cell lines having an IC_{50} value less than 2.68 μM. The introduction of the *N*-benzyl group at the indole nitrogen also increased the activity [37].

2.2 Benzimidazolium salts having aromatic and aliphatic substituents

Akkoc et al. reported benzimidazolium salts screened for anticancer potential against human embryonic kidney (HEK-293 T), human colon epithelial colorectal adenocarcinoma (DLD-1), and human breast epithelial adenocarcinoma (MDA-MB-231) cancer cell lines by using MTT assay. Palladium (Pd) metal complexes were also prepared and found inactive against these cells lines having IC_{50} values over 100 μM. The naphthalen-1-yl-methyl incorporated benzimidazolium chloride **15** (**Figure 4**) (IC_{50} = 26.09 μM) showed most cytotoxicity against DLD-1 cell line [38]. In an additional study, Akkoc and his coworkers reported a

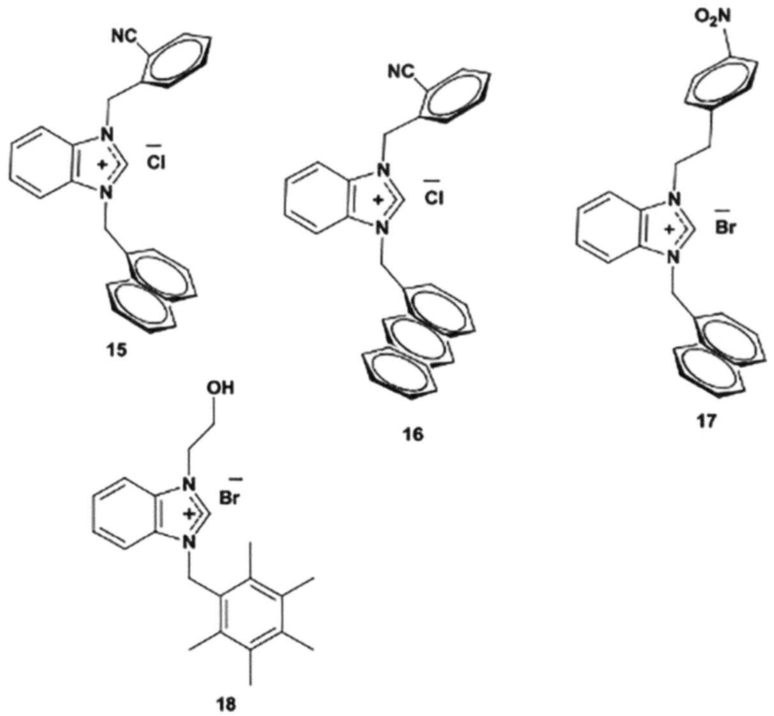

Figure 4.
Benzimidazole salts having aromatic substituents.

novel series of benzimidazolium salts containing cyanobenzyl, nitrophenyl, and *N*-methylphthalimide substitutions. All salts were tested for their cytotoxicity against DLD-1 and human breast cancer (MDA-MB-231) cell lines. Compounds **16** and **17** demonstrated exceptional activity against MDA-MB-231 cell lines with IC_{50} values of 1.26 and 2.01 μM and were considerably better than cisplatin (IC_{50} = 5.77 μM). Derivatives **16** and **17** contain anthracene and naphthalene rings, respectively, attached with the nitrogen of benzimidazole. The N^1 and N^3 substituents produced a prominent effect on anticancer activity [39]. In 2019, Akkoc extended his previous finding of cytotoxicity of benzimidazolium salts. In this regard, 2-hydroxyethyl-containing benzimidazolium salts along with respective Pd-complexes were prepared and their anticancer capacity was noted against human cancer cell lines. Surprisingly, compound **18** exhibited notable activity against MDA-MB-231 (IC_{50} = 7.59 μM) and DLD-1 (IC_{50} = 39.51 μM) cell lines in comparison with their Pd-complexes [40].

Lin et al. synthesized 1,3-bis-naphthyl-substituted benzimidazolium bromides and estimated for activity against MDA-MB-468 as well as PC-3 cell lines. As compared to the standard drug tamoxifen (IC_{50} = 22.5 μM), compound **19** (**Figure 5**) was found as an active agent (IC_{50} = 9.7 μM) against MDA-MB-468. The presence of the naphthyl group was vital for the activity of these derivatives [41]. Wright et al. synthesized naphthalene-substituted imidazolium salts and evaluated the anticancer performance against non-small-cell lung cancer (NSCLC) cell lines. The anticancer activity was evaluated by the MTT assay. Compound **20**, the benzimidazolium salt, displayed IC_{50} values of 3, 4, and 5 μM against NCI-H460, NCI-H1975, and HCC-827, respectively. Compound **20** carries naphthyl rings at both nitrogens of benzimidazole [42]. Stromyer et al. synthesized benzimidazole

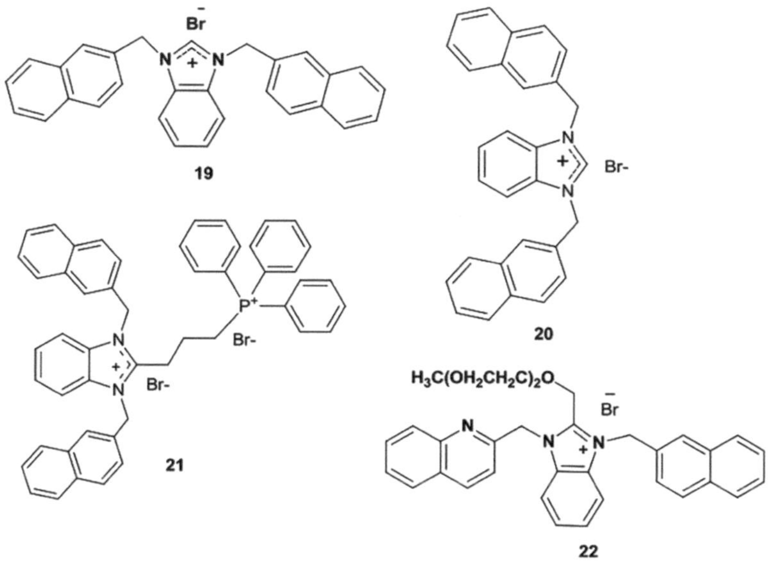

Figure 5.
Benzimidazole salts having naphthalene and quinoline rings.

salts having triphenylphosphonium group. The nitrogen atoms of benzimidazole were linked with naphthyl methyl groups. Compound **21** revealed marvelous activity against bladder cancer cell lines RT4, RT112, UMUC3, and SW780. This compound also showed apoptosis by causing mitochondrial damage. The drug causes a rapid and irreversible effect against bladder cancer [43]. Shelton et al. synthesized *N,N*-bis-arylmethyl-substituted benzimidazolium salts *via* cyclization of *o*-phenylenediamine or 2-(2-(2-methoxy ethoxy)ethoxy)acetic acid with 2-(chloromethyl) quinolone or 2-(bromomethyl)-naphthalene followed by alkylation and quaternization. Various hydrophilic and hydrophobic groups were added at both nitrogen atoms of benzimidazole. Insight into *in vitro* cytotoxicity of synthesized salts, compound **22** showed adequate activity against NSCLC cancer cells having IC_{50} values ranging between 1 and 7 μM comparable to the standard drug, cisplatin. This compound bears quinoline and naphthalene rings to both nitrogen atoms of benzimidazole. The presence of ether linkage at position two increased the hydrophilicity of this compound [44].

Bansode et al. carried out the synthesis of ferrocene-linked ionic liquids by incorporating long alkyl chains. Anticancer activity was evaluated against MCF-7 by using sulforhodamine B assay. These ferrocene-quaternized azolium salts showed significant cytotoxic potential against MCF-7 and 1-(ferrocenylmethyl)-3-tetradecylbenzimidazolium bromide **23** (**Figure 6**) was found to be most potent (GI_{50} = 0.016 μM) as compared to standard drug doxorubicin (GI_{50} = 0.018 μM). Derivatives in this followed the Lipinski rule of five and showed excellent pharmacokinetic properties [45]. Kucukbay et al. synthesized *N, N*-disubstituted benzimidazolium bromides and evaluated anticancer activity against PC-3 and ovarian (A2780) cancer cell lines. Derivatives **24-26** presented prominent activity against PC-3 and A2780 cancer cell lines having IC_{50} values in micromolar concentration. These compounds bear a 4-methoxyphenyl ethyl group at the benzimidazole nitrogen [46]. Haque et al. prepared a collection of bis-benzimidazolium salts and evaluated against human colon cancer (HCT-116). All compounds showed superior activity than the reference

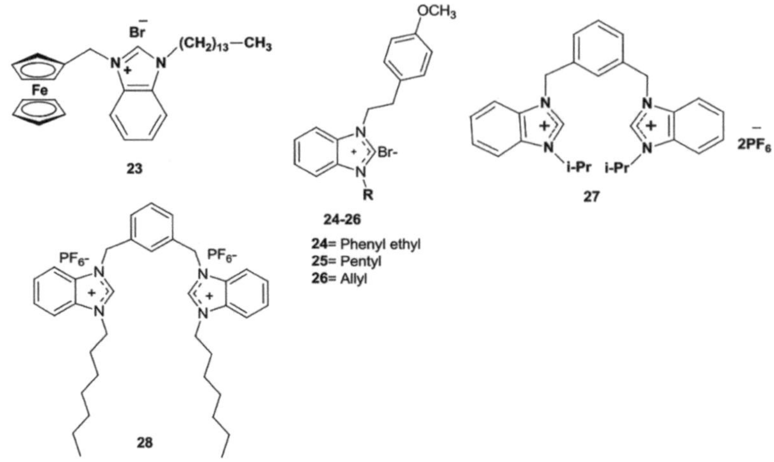

Figure 6.
Bis-benzimidazolium and ferrocene-linked benzimidazolium salts.

drug, fluorouracil. Derivatives having *N*-methylene phenyl substituents presented prominent activities. The highest anticancer potential was observed in the case of derivative **27** (IC_{50} = 0.2 μM) remarkably better than reference drug, 5-fluorouracil (IC_{50} = 19.2 μM) [47]. Noor ul Huda et al. synthesized bis-NHC benzimidazolium salts and evaluated them as antimicrobial and anticancer agents. Compound **28** exhibited prominent activity against HCT-116 cancer cell lines showing 75% inhibition at 1 mg/ml as determined by using sulforhodamine B assay. Compound **28** contains a lipophilic alkyl chain and the lipophilicity of the alkyl chain was linked to the increased activity of this derivative [48].

2.3 Benzimidazolium silver metal complexes

Cisplatin is the first metal-based drug used for the cure of cancer [49]. The serendipitous discovery of cisplatin stimulated the search for new metal-based anticancer agents. Silver (Ag) salts have been used as antimicrobial agents for purification of drinking water and wound healing [50, 51]. Based on its antimicrobial property, silver has also been explored as an anticancer agent. *N*-Heterocyclic carbenes (NHC) are a prominent family of organometallic ligands.

2.3.1 Bis-benzimidazolium silver metal complexes

Iqbal and his coworkers performed a detailed study to reduce the risk of malignant neoplasm and reported novel binuclear benzimidazolium salt and corresponding Ag (I) NHC complex. Compound **29** (**Figure 7**)presented prominent activity (IC_{50} = 1.7 μM) against HCT-116 cell line. This compound also showed significant inhibition of inflammatory cytokines such as tumor necrosis factor-alpha and interleukin in human macrophages. Compound **29** showed apoptotic activity *via* inhibition of the caspase pathway. Photomicrographs of the cell treated with compound **29** showed deposition of silver in cells [52]. Gadhayeb et al. carried out the synthesis of mono- and bis-NHC complexes having palladium (Pd) and silver metals. The anticancer activity was evaluated out against HCT-116 cell line. Compounds

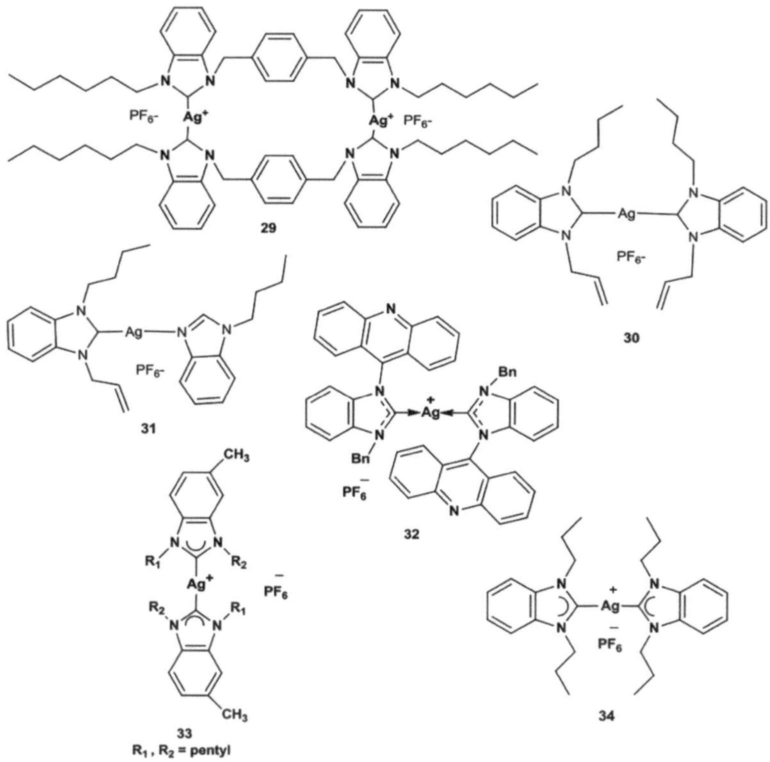

Figure 7.
Mono and binuclear bis-benzimidazolium silver metal complexes.

30 and **31** exhibited prominent activities having IC_{50} values of 12.3 ± 1.2 μM and 10.6 ± 1.8 μM, respectively. The mono-NHC showed more activity as compared to bis-NHC. The increased activity of compound **31** could be due to the increased release of silver from the mono-NHC complex. These compounds contain a butyl chain at the benzimidazole nitrogen [53]. Following this principle, Sarhan et al. reported benzimidazolium-acridine-based salts and metal complexes with pronounced biological potential. Specifically, compound **32** can be considered an excellent *in vitro* anticancer agent against MCF-7 (IC_{50} = 21 μM) and selectivity index of 3.6. Therefore, Ag-NHC complexes demonstrated prominent activity [54]. A series of 5-methyl benzimidazole-based *n*-heterocyclic carbene (NHC) salts and their silver (I)-complexes were prepared by Habib et al. The Ag (I)-benzimidazolium complexes showed dominant activity against human breast cancer (MDA-MB-231) and colon cancer (HCT-116) as compared to NHC salts. Compound **33** showed promising activity against MDA-MB-231 (IC_{50} = 4.2 ± 0.24 μM, SI = 7.63) and HCT-116 (IC_{50} = 7.43 ± 0.23 μM SI = 4.33) as compared to reference drug (IC_{50} = 8.20 ± 0.14 μM and 5.5 ± 0.34 μM) against these cell lines respectively. Compound **33** carries pentyl chains at both nitrogen atoms of the benzimidazole core. Derivatives having a longer alkyl chain were found more active. These molecules showed dose-dependent cytotoxicities and apoptosis by mitochondrial pathways [55]. A range of substituted Ag(I)-benzimidazolium carbene complexes were reported by Atif et al. and *in vitro* anticancer studies were carried out against MCF-7, HCT 116, and erythromyeloblastoid leukemia (K-562) cell lines. Promising anticancer activity was shown by **34** (IC_{50} = 0.31 μM) against the K-562 cell line. Compound **34** is a bis-benzimidazole silver complex and carries propyl groups at both nitrogen atoms of benzimidazole rings [56].

Achar et al. reported a novel series based on benzimidazolium salts linked with coumarin heterocycle, and their silver cationic bis-NHC and Ag neutral mono-NHC were synthesized. The anticancer activity was evaluated by sulforhodamine assay. Complex **35** (**Figure 8**) exhibited moderate activity against the A549 cell line having an IC_{50} value of 8.3 ± 0.40 μM. Compound **35** is the bis-NHC-coordinated Ag hexafluorophosphate (PF6) salt. The mono-NHC coordinates Ag acetate complexes showed inferior activity with IC_{50} values greater than 10 μM. Bis-NHC complexes also showed prominent antibacterial activity [57]. Yasar et al. also worked out to yield novel zwitterionic-sulfonated benzimidazolium salts and their Ag-(I) complexes by following a reported synthetic approach. Anticancer activity was assessed against human cervix carcinoma (HeLa), human adenocarcinoma (HT29), and mouse fibroblast (L929) cancer cell lines. Silver complexes were found most active as compared to their salts. Compound **36** (IC_{50} = 11 ± 1 μM) showed higher potency against HT29 cancer cell line as compared to cisplatin (IC_{50} = 42 ± 6 μM). Compound **36** was found to be the least toxic (IC_{50} = 126 ± 3 μM) against non-cancer L929 cell lines [58]. Karlık et al. prepared a series of aqua-bis-benzimidazole Ag (I) *p*-toluene sulfonate complexes *via* a multistep approach and investigated their anticancer properties against human colorectal (Caco-2) and MCF-7 cancer cell lines. Salts were found to be ineffective against these cell lines. Benzimidazolium Ag complex **37** (IC_{50} value of 9 ± 3 μM) showed excellent activity against the Caco-2 cell line but was inactive against the MCF-7 cell line. Derivatives **37** carries an *o*-chloro-substituted benzyl group attached with the nitrogen of benzimidazole and this substituent was found more effective at

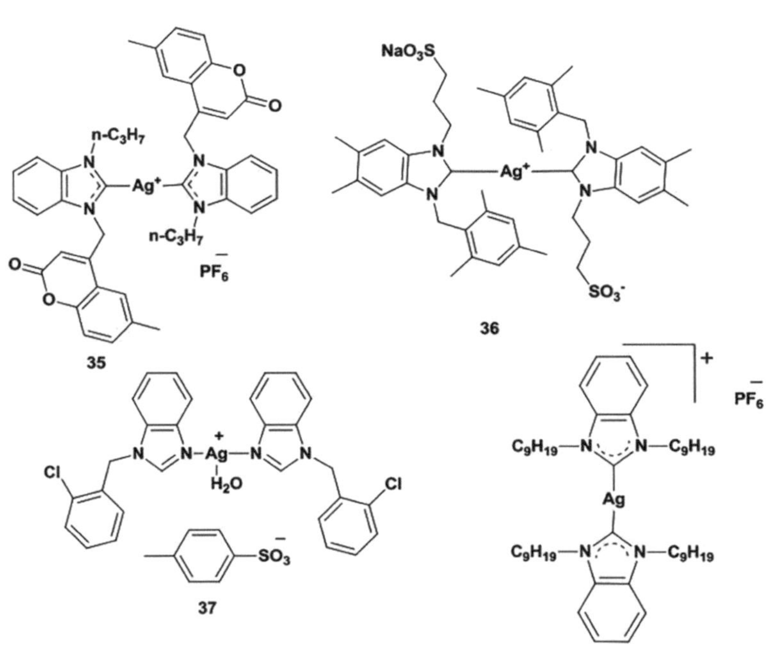

Figure 8.
Bis-benzimidazolium silver metal complexes having alkyl chain, heterocyclic and aromatic scaffold.

this position as compared to *o*-methyl and *p*-methyl analogues [59]. Fatima et al. successfully explored the cytotoxic effect of different alkyl chains on benzimidazolium-based Ag complexes. It was observed that compound **38** containing longer n-alkyl chains showed the best cytotoxic potential against HCT-116 (IC_{50} = 0.02 μM) as compared to 5-fluorouracil (IC_{50} = 10.2 μM) as a standard drug. The incorporation of silver ions and elongation of the alkyl chain amplified the anticancer activity [60].

Similarly, Haque et al. prepared Ag(I) complexes containing nitrile-functionalized benzimidazolium salt as an active agent against HCT-116. Among all synthesized complexes, compound **39** (IC_{50} = 14.9 ± 0.8 μM) showed the highest cytotoxicity as compared to fluorouracil (IC_{50} = 5.2 ± 0.3 μM). The activity was linked to the nitrile group (**Figure 9**) at the meta position of the benzyl group causing the weak electron-withdrawing effect [61]. Early on, Haque et al. also synthesized silver metal complexes containing benzimidazolium ligand. These silver-based benzimidazolium complexes are capable of slow release of the silver ion at the cancerous cell, affecting cell morphology. Complex **40** (IC_{50} = 1.20 ± 0.3 μM), a binuclear silver entity, expressed superior activity as compared to fluorouracil (IC_{50} = 5.2 ± 0.3 μM) against HCT-116. Complex **40** was found to be the least active (IC_{50} = 103 ± 2.3 μM) against the HT29 cell line. Therefore, the HT-29 cell line was found to be resistant to this complex [62]. Hussaini et al. successfully formulated silver (I)-benzimidazolium carbenes. A series of propylene-linked bis-benzimidazolium salts having different alkyl chains and respective binuclear silver complexes were prepared. Complexes showed dose-dependent cytotoxicities. Their cytotoxic studies against the MCF-7 cell line revealed compound **41** as the most active (IC_{50} = 7 ± 1 μM) complex as compared to tamoxifen (IC_{50} = 11 ± 2 μM). The presence of the propyl chain at the benzimidazole nucleus was found to be optimum for anticancer activity [63]. Later on, in 2018,

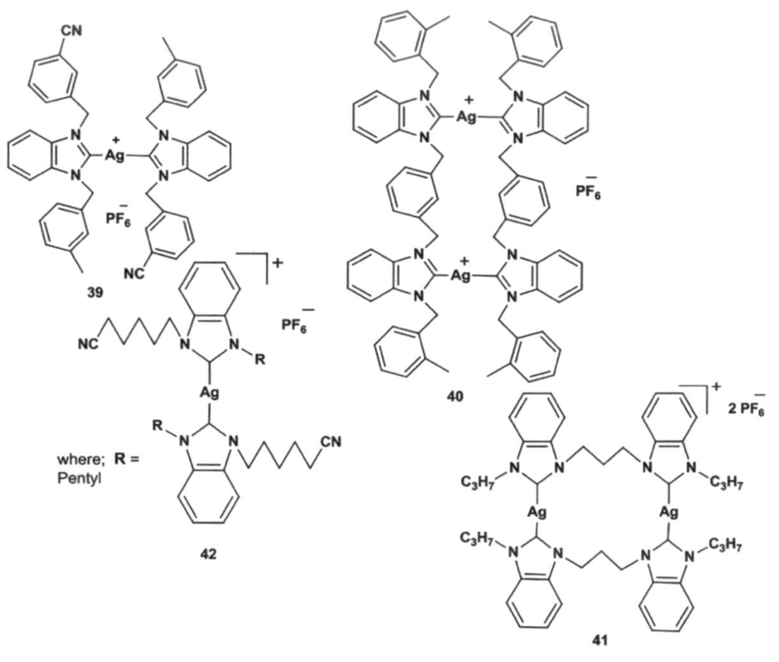

Figure 9.
Bis-benzimidazolium silver metal complexes containing aliphatic nitriles, aromatic nitriles, benzyl and alkyl chains.

Hussaini and his coworkers reported benzimidazolium salts having aliphatic nitrile group and their Ag (I)-benzimidazolium carbenes complexes. All of these NHC complexes showed good cytotoxicities with IC_{50} values in the range of 7.0–12.9 μM against the MCF-7 cell line. Compound **42** (IC_{50} = 7 ± 1.06 μM) was found to be the most potent and it carries a pentyl chain attached to the benzimidazole nitrogen. Cytotoxicities of these compounds increase as the alkyl chain length expands [64].

2.3.2 Benzimidazolium silver and gold metal complexes

Akkoc et al. reported a series of silver- and palladium-based metal complexes with benzimidazolium ligand. This attempt was made in search of the non-platinum antitumor drugs due to the observed side effects of cisplatin and nedaplatin. However, silver complex **43** (**Figure 10**) showed promising *in vitro* cytotoxic potential against DLD-1 (IC_{50} = 12.41 μM), MDA-MB-231 (IC_{50} = 11.98 μM) cancer cell lines, and HEK-239 (IC_{50} = 4.2 μM) non-cancer cell lines. Therefore, the silver complex was found more potent than the palladium complex [65]. Sahin et al. employed the conventional technique to synthesize *n*-allyl-substituted benzimidazolium-based carbene and corresponding Ag-(I) complexes. The first step was to obtain *n*-alkylated benzimidazole and later followed by salt formation. Silver dioxide was used to obtain the allyl-linked benzimidazolium Ag-(I) complex. Among all synthesized compounds, derivatives **44** (IC_{50} = 1.41 μM) and **45** (IC_{50} = 1.21 μM) showed the best *in vitro* anticancer potential against DU-145 and MCF-7, and MDA-MB-231 (IC_{50} < 1 μM) cancer cell lines. Derivatives **44** and **45** also exhibited some degree of selectivity [66]. Ozdemir et al. carried out the synthesis of silver and gold (Au) NHC-propyl sulfonate complexes. Silver complexes presented dominant activity as compared to gold salts. Compounds **46** (IC_{50} = 2.32 ± 0.089 μM) and **47** (IC_{50} = 9.31 ± 0.95 μM) presented prominent *in vitro* activities against adenocarcinoma (HEP3B) cancer cell lines. Complexes **46** and **47** are silver and gold complexes, respectively. These complexes

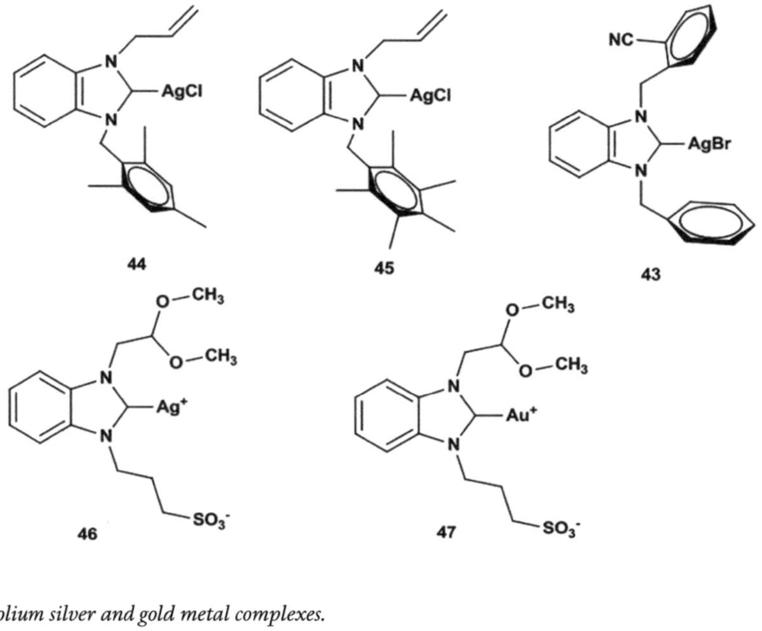

Figure 10.
Benzimidazolium silver and gold metal complexes.

possess a dimethoxyethyl group at the benzimidazole nitrogen. Replacement of dimethoxyethyl group by diethoxyethyl group produced less active derivatives [67].

Similarly, the synthetic work of Cevik Yildiz and his coworkers resulted in the formation of novel benzimidazolium salt. These benzimidazolium salts were used as a ligand to obtain corresponding novel Ag (I)- benzimidazolium complexes. Cytotoxic studies of salts and complexes were carried out against MCF-7, MDA-MB-23, DU-145 by MTT assay. Compound **48** (**Figure 11**) displayed prominent activity (IC_{50} < 1 μM) against breast cancer MCF-7 and MDA-MB-23 cell lines. Compound **49** also showed prominent activity (IC_{50} < 1 μM) against the MCF-7 cell line than standard drug. These compounds showed concentration-dependent killing and also showed selectivity for cancer cell lines [68]. Aktas et al. carried out the synthesis of 2-morpholine ethyl-substituted benzimidazolium salts and their Ag-NHC complexes. Compounds **50** (IC_{50} = 6.59 μM) and **51** (IC_{50} = 6.56 μM) exhibited prominent activity against MCF-7 cell line. Ag-NHC complexes presented prominent activity as compared to benzimidazolium salts. Compounds **50** and **51** carry methyl and tetra-methyl benzyl groups at the benzimidazole nitrogen [69]. A series of di-isopropylamine ethyl benzimidazolium salts have been reported by Kızrak et al. These salts were further used as a precursor for the syntheses of corresponding silver and gold benzimidazolium carbene complexes. Resultant derivatives displayed prominent activity against the human brain (SHSY5Y) cell line and compounds **52, 53** were most prominent presenting IC_{50} values of 5.23 and 4.74 μM, respectively. Compounds **52** and **53** are

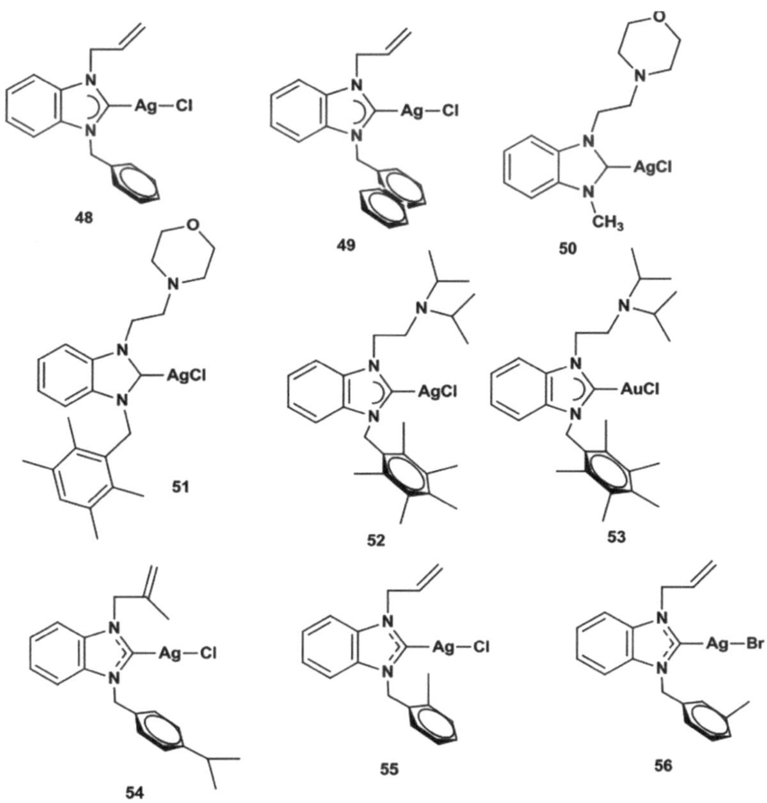

Figure 11.
Benzimidazolium silver metal complexes.

silver and gold complexes, respectively. Therefore, the gold complex was found more potent than silver complexes. Compound **52** was also found significant against HEP3B (IC_{50} = 6.19 ± 1.09 μM) and HTC-116 (IC_{50} = 8.44 ± 1.07 μM) cancer cell line [70]. Sahin-Bolukbasi and Sahin synthesized two new benzimidazolium salts and reacted with silver dioxide (Ag_2O) to obtain Ag-(I) benzimidazolium complexes. Evaluation of the anticancer potential of all these compounds reflected the higher cytotoxicity (IC_{50} < 1 μM) of **54** against DU-145, MCF-7, and MDA-MB-231. This compound carries 2-methyl propenyl and *p*-isopropyl benzyl substituents at the N^1 and N^3 positions of benzimidazole ring [71]. Early on, Sahin-Bolukbasi et al. synthesized unsymmetrical benzimidazolium salt and respective Ag-(I) complexes. All of the synthesized compounds were subjected to cytotoxic evaluation. Higher cytotoxicity was noted for benzimidazolium-based Ag complexes as compared to benzimidazolium salts, and particularly **55** and **56** were most active (IC_{50} < 1 μM) against MCF-7, MDA-MB-231 cancer cell lines. These compounds also demonstrated activity against DU-145 cell line having IC_{50} values of 6.02 ± 0.30 μM and 5.16 ± 0.33 μM, respectively. The *ortho*-substituted benzyl group proved more active as compared to *meta*- and *para*-substituted benzyl groups. And in *ortho*-substituted derivatives methyl group was found more potent as compared to the chlorine atom [72].

2.4 Selenium-based benzimidazolium salts and complexes

Selenium (Se) is very important for the human body and its deficiency can lead to cancer, diabetes, and cardiovascular diseases [73]. Selenium is present in some food and drinks in traces [74]. Recently, selenium has been incorporated in new anticancer agents due to its low toxicity. Kamal et al. applied a green synthetic approach to obtain novel benzimidazolium salts and Se-based benzimidazolium-heterocyclic carbenes. The *in vitro* anticancer potential was evaluated against RGC-5, Hela, MCF-7, and mouse melanoma (B16F10) cancer cell lines using fluorouracil as a standard drug. Prominent anticancer activity was observed against Hela and RCG cell lines. Among benzimidazolium salts (**Figure 12**), compounds **57** (IC_{50} = 0.04 ± 0.31 μM) and **58** (IC_{50} = 0.24 ± 0.22 μM) displayed prominent activity against Hela cell line.

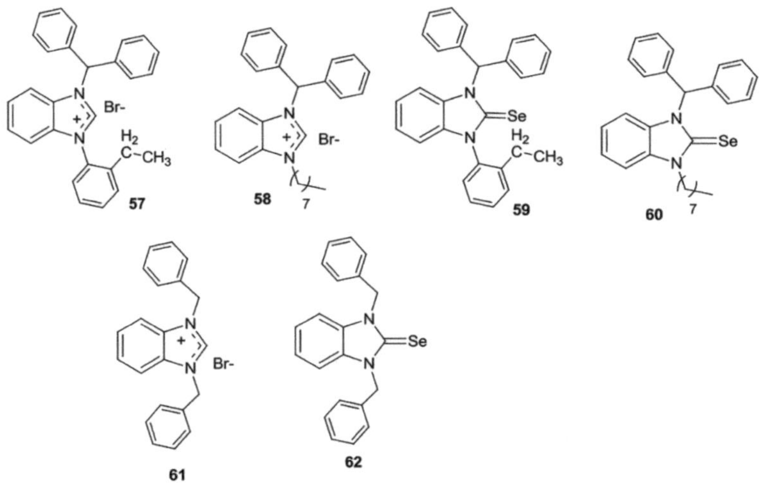

Figure 12.
Benzimidazolium salts and selenium metal complexes.

The corresponding Se-NHC-adducts compounds **59** (IC_{50} = 0.11 ± 0.20 μM) and **60** (IC_{50} = 4.3 ± 0.11 μM) were found most active against Hela cell line. Compound **59** was also effective against RCG cell line (IC_{50} = 9.16 ± 0.27 μM). Molecular docking investigation of compounds **59** and **60** with epidermal growth factor (EGF), vascular endothelial growth factor (VEGF), and cyclooxygenase (COX-1) displayed strong interactions [75]. Similarly, green synthesis of benzimidazolium salts and di-Se-*N*-heterocyclic carbene complexes was carried out by Iqbal et al. The benzimidazolium salts **61** (IC_{50} = 3.94 μM) and respective Se-NHC **62** (IC_{50} = 3.49 μM) displayed prominent activity against HCT-116 cell line as compared to fluorouracil (IC_{50} = 4.9 μM). Both compounds possess benzyl groups at positions 1 and 3 of the benzimidazole nucleus. Compounds **61** and **62** also showed some degree of apoptosis by the mitochondrial pathway. Further pro-apoptotic evaluation for HCT-116 even at low concentration due to a strong release of selenium metal for DNA interaction was successfully reported [76].

2.5 Ruthenium complexes based on benzimidazolium salts

Organic compounds having ruthenium (Ru) metal are also being used as anticancer agents. Akkoc et al. synthesized methylpyridine-linked benzimidazolium salt and corresponding Ru (II) complexes containing benzimidazolium ligand. The antiproliferative assay revealed that **63** (**Figure 13**) showed DNA binding as well as the best cytotoxic potential against MCF-7 (IC_{50} = 23.8 μM), Caco-2 (IC_{50} = 18.0 μM) cell lines, respectively. Compound **63** carries a hexamethyl phenyl ring at position 2 of the benzimidazole ring. Compound **63** showed electrostatic and hydrophobic interaction with DNA and presented a binding affinity of −13.779 kcal/mol [77]. Omar et al. synthesized benzotriazole-functionalized palladium and ruthenium complexes. Comparison of cytotoxic studies revealed that ruthenium complexes are better than palladium complexes against MCF-7 and Caco-2 cancer cell lines. Although the activity displayed by these compounds

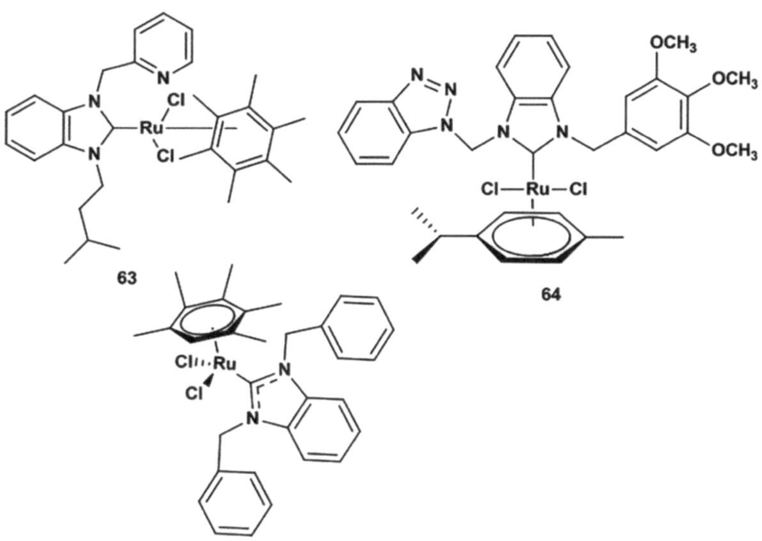

Figure 13.
Ruthenium complexes.

was not prominent, these compounds were found inert for normal cell line L-929. Compound **64** (IC_{50} = 90 μM) is one member of this series [78]. Lam and his coworkers aimed for the novel benzimidazolium-based Ru complexes with halogen ligands for exploration of cytotoxic potential against HCT-116, SiHa, and NCI-H460 cell lines. Promising cytotoxic potential was observed in case of **65** against HCT-116 (IC_{50} = 6.2 ± 0.4 μM), SiHa (IC_{50} = 8.4 ± 0.2 μM), and NCI-H460 (IC_{50} = 7.8 ± 1.0 μM) cancer cell lines. Replacement of ruthenium by osmium (Os) also presented equal cytotoxicity against HCT-116 cell line [79].

2.6 Miscellaneous metal complexes

As mentioned earlier, NHC-carbenes are highly reactive species depending upon the nature of the ligand and transition metal used; consequently, different transition metals have been coordinated with benzimidazolium ligand for a better cytotoxic effect. Troung et al. studied the cytotoxicity of rhodium (Rh)- and iridium (Ir)-based benzimidazolium complexes. A series of rhodium and iridium complexes were prepared and evaluated for anticancer potential against HCT-116, NCI-H460, SiHa, SW480 human cancer cell lines. Compounds **66** and **67** (**Figure 14**)

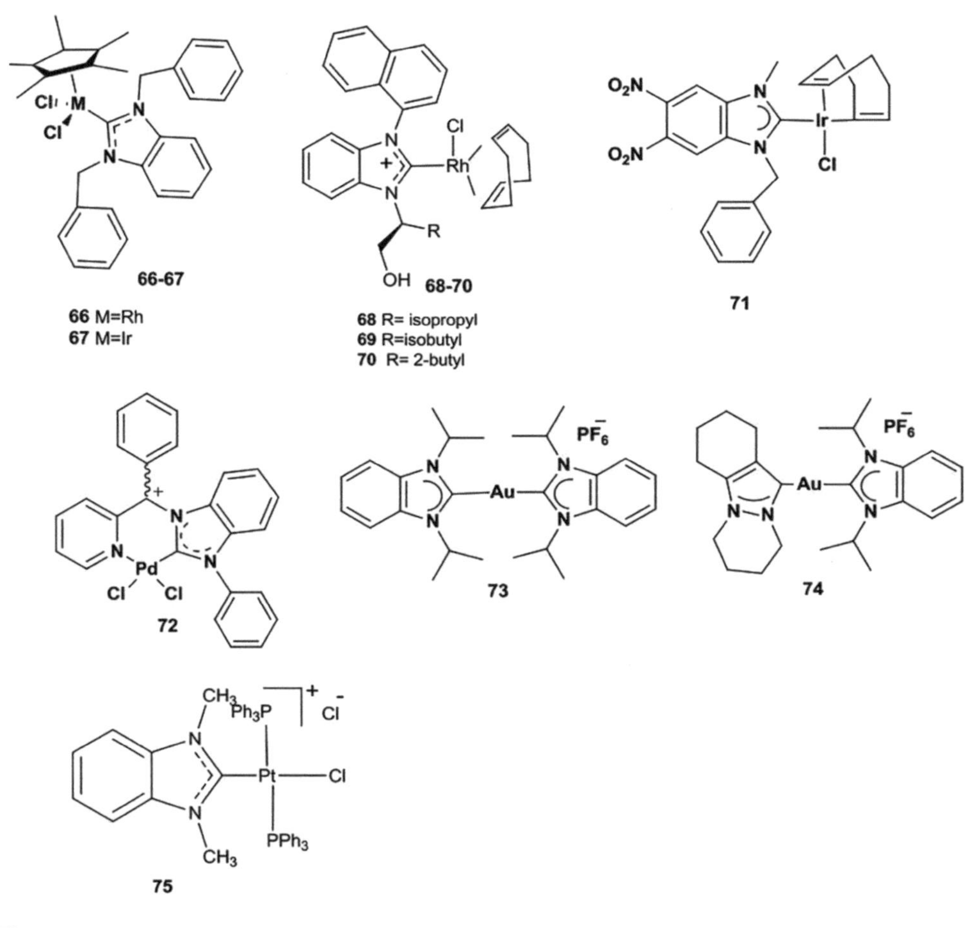

Figure 14.
Benzimidazole metal complexes with gold, rhodium, iridium, and palladium.

advertised notable activity against HCT-116 (IC_{50} = 7.4 ± 0.5 μM, 11 ± 0.1 μM), NCI-H460 (IC_{50} = 11 ± 1 μM, 23 ± 3 μM), SiHa (IC_{50} = 10 ± 1 μM, 19 ± 1 μM), SW (IC_{50} = 5.8 ± 1 μM, 19 ± 1 μM) cancer cell lines. Compounds **66** and **67** are Ir and Rh metal complexes, respectively. These compounds contain benzyl groups at both nitrogen atoms of benzimidazole. Studies of the mode of action of rhodium complexes are based on the fact that instead of interaction with DNA, it accumulates in the cytoplasm [80]. Zhao et al. synthesized naphthyl NHC-Rh complexes by incorporating a hydroxy alkyl chain at benzimidazole nitrogen. Upon evaluation against MCF-7 cell line, compounds **68** (IC_{50} = 0.38 μM), **69** (IC_{50} = 0.45 μM), and **70** (IC_{50} = 0.72 μM) showed excellent activity as compared to standard drug paclitaxel (IC_{50} = 1.38 μM). Therefore, the introduction of the hydroxyl group and alkyl group at the benzimidazole nitrogen is beneficial for the activity [81]. Sanchez-Mora et al. reported the formation of two benzimidazolium-based Ir(I) complexes as new cytotoxic agents. The benzimidazole ring was substituted by benzyl and pentafluorobenzyl groups. The benzyl-substituted derivative presented prominent activity and compound **71** was found to be the strongest agent against PC-3 (IC_{50} = 10.6 ± 0.9 μM) and SKLU-7 (IC_{50} = 10.4 ± 1.5 μM) cell lines. This compound showed less toxicity for normal cell line COS-7 [82]. Choo et al. synthesized pyridine-functionalized Pd-based imidazolium and benzimidazolium carbenes. Imidazolium Pd complex showed prominent activity against cancer cell lines. Benzimidazolium Pd complex **72** is also an excellent candidate for anticancer studies. Generally, NHC-Pd complexes create covalent bonding with DNA resulting in cross-linking of guanine base [83]. Early on in 2012, Sivaram and his coworker were able to synthesize new gold (I) and gold (III) complexes bearing benzimidazolium ligand. These complexes were mono-, homo-bis-, and hetero-bis-benzimidazole NHC. The hetero-bis-benzimidazole complexes are nonclassical pyrazole-derived NHC. Complexes **73** (IC_{50} = 0.284 ± 0.11 μM) and **74** (IC_{50} = 0.24 ± 0.01 μM) exhibited prominent inhibitory action against NSCLC (NCI-H1666) cell line. These complexes are isopropyl-substituted homo-bis and hetero-bis NHC complexes, respectively [84]. Rehm et al. synthesized benzimidazole platinum complexes having different alkyl chains such as methyl, ethyl, butyl, and octyl chains. The bisphosphane complexes showed the excellent anticancer activity against seven cancer cell lines. Compound **75** appeared as the most effective against different cancer cell lines having IC_{50} values placing from 0.10 to 0.30 μM. Complex **75** displayed prominent activity (IC_{50} = 0.10 ± 0.01 μM) against multidrug-resistant strains of MCF-7. Cell cycle analysis of some complexes indicated that they produced cell blockage in the G1 stage [85].

3. Conclusion

The pharmacological properties of benzimidazolium salts have attracted the attention of medicinal chemists. The resemblance of benzimidazole scaffold with purine bases establishes it biologically significant. Benzimidazolium salts have demonstrated promising activities against various cancer cell lines. Benzimidazolium salts derivatives have been prepared by the functionalization of two nitrogen atoms in the imidazole ring along with the preparation of hybrid molecules, and metal complexes. The hybridization of benzimidazole salts with natural compounds such as chalcones and steroids exhibited prominent activities (IC_{50} < 1 μM) against various cancer cell lines. Some hybrids compounds also showed the phenomenon of apoptosis. Compounds carrying alkyl chains and aromatic rings at benzimidazole nitrogen showed pronounced activity. The introduction of phenyl, naphthalene, anthracene, and quinoline rings at the benzimidazole nitrogens through methylene groups intensified the anticancer activity. The 5,6-dimethyl-substituted benzimidazole derivatives were also found more active as compared to unsubstituted benzimidazole rings. In the case of silver metal complexes, bis-benzimidazolium complexes exhibited exceptional activity against colon cancer ($IC_{50} < 1\ \mu M$) cell line. But silver metal complexes presented less selectivity indices. Mono-benzimidazolium metal complexes proved more active against breast cancer ($IC_{50} < 1\ \mu M$) cell lines. In some derivatives, the introduction of a long alkyl chain at the benzimidazole nitrogen is beneficial for the augmentation of anticancer activity. The activity of selenium metal complexes was almost equivalent to their respective salts. While the halogen-substituted ruthenium benzimidazole metal complexes showed moderate activity. Rhodium, platinum, and gold complexes have also shown encouraging anticancer activities and are excellent candidates for future investigations. The *in vivo* investigations of potent compounds mentioned in this chapter could lead to further developments in the field.

Author details

Imran Ahmad Khan[1], Noor ul Amin Mohsin[2], Sana Aslam[3] and Matloob Ahmad[1*]

1 Department of Chemistry, Government College University, Faisalabad, Pakistan

2 Department of Pharmaceutical Chemistry, Faculty of Pharmaceutical Sciences, Government College University, Faisalabad, Pakistan

3 Department of Chemistry, Government College Women University Faisalabad, Pakistan

*Address all correspondence to: matloob.ahmad@gcuf.edu.pk

References

[1] Bray F, Ferlay J, Soerjomataram I. Global cancer statistics 2018: GLOBOCAN estimates of incidence and mortality worldwide for 36 cancers in 185 countries. CA: A Cancer Journal for Clinicians. 2018;**68**(6):394-424. DOI: 10.3322/caac.21492

[2] Bray F, Jemal A, Grey N. Global cancer transitions according to the Human Development Index (2008-2030): A population-based study. The Lancet Oncology. 2012;**13**(8):790-801. DOI: 10.1016/S1470-2045(12)70211-5

[3] Danaei G, Vander Hoorn S, Lopez AD, Comparative Risk Assessment collaborating group (Cancers). Causes of cancer in the world: Comparative risk assessment of nine behavioural and environmental risk factors. The Lancet. 2005;**366**(9499):1784-1793. DOI: 10.1016/S0140-6736(05)67725-2

[4] Pucci C, Martinelli C, Ciofani G. Innovative approaches for cancer treatment: Current perspectives and new challenges. Ecancermedicalscience. 2019;**13**:961. DOI: 10.3332/ecancer.2019.961

[5] Masoud V, Pages G. Targeted therapies in breast cancer: New challenges to fight against resistance. World Journal of Clinical Oncology. 2017;**8**(2):120-134. DOI: 10.5306%2Fwjco.v8.i2.120

[6] Vitaku E, Smith DT, Njardarson JT. Analysis of the structural diversity, substitution patterns, and frequency of nitrogen heterocycles among us FDA approved pharmaceuticals. Journal of Medicinal Chemistry. 2014;**57**(24): 10257-10274. DOI: 10.1021/jm501100b

[7] Barker HA, Smyth RD, Weissbach H. Isolation and properties of crystalline cobamide coenzymes containing benzimidazole or 5, 6-dimethy lbenzimidazole. Journal of Biological Chemistry. 1960;**235**(2):480-488. DOI: 10.1016/s0021-9258(18)69550-x

[8] Khan MT, Razi MT, Jan SU. Synthesis, characterization, and antihypertensive activity of 2-phenyl substituted benzimidazoles. Pakistan Journal of Pharmaceutical Sciences. 2018;**31**(3):1067-1074. DOI: 10.36721/pjps.2019.32.6.reg.2585-2597.1

[9] Kaur G, Silakari O. Benzimidazole scaffold-based hybrid molecules for various inflammatory targets: Synthesis and evaluation. Bioorganic Chemistry. 2018;**80**:24-35. DOI: 10.1016/j.bioorg.2018.05.014

[10] Al-Blewi FF, Almehmadi MA, Aouad MR. Design, synthesis, ADME prediction, and pharmacological evaluation of novel benzimidazole-1, 2, 3-triazole-sulfonamide hybrids as antimicrobial and antiproliferative agents. Chemistry Central Journal. 2018;**12**(1):1-14. DOI: 10.1186/s13065-018-0479-1

[11] Scheinfeld N. Cimetidine: A review of the recent developments and reports in cutaneous medicine. Dermatology Online Journal. 2003;**9**(2):4. DOI: 10.5070/D33s15q645

[12] Vausselin T, Seron K, Lavie M. Identification of a new benzimidazole derivative as an antiviral against hepatitis C virus. Journal of Virology. 2016;**90**(19): 8422-8434. DOI: 10.1128/JVI.00404-16

[13] Ozil M, Parlak C, Baltas N. A simple and efficient synthesis of benzimidazoles containing piperazine or morpholine skeleton at C-6 position as glucosidase

inhibitors with antioxidant activity. Bioorganic Chemistry. 2018;**76**:468-477. DOI: 10.1016/j.bioorg.2017.12.019

[14] Hussain A, Alajmi MF, Rehman M. Evaluation of transition metal complexes of the benzimidazole-derived scaffold as promising anticancer chemotherapeutics. Molecules. 2018;**23**(5):1232-1250. DOI: 10.3390/molecules23051232

[15] Bharadwaj SS, Poojary B, Nandish SKM. Efficient synthesis and in silico studies of the benzimidazole hybrid scaffold with the quinolinyloxadiazole skeleton with potential α-glucosidase inhibitory, anticoagulant, and antiplatelet activities for type-II diabetes mellitus management and treating thrombotic disorders. ACS Omega. 2018;**3**(10):12562-12574. DOI: 10.1021/acsomega.8b01476

[16] Wang YT, Shi TQ, Zhu HL. Synthesis, biological evaluation, and molecular docking of benzimidazole grafted benzsulfamide-containing pyrazole ring derivatives as novel tubulin polymerization inhibitors. Bioorganic & Medicinal Chemistry. 2019;**27**(3):502-515. DOI: 10.1016/j.bmc.2018.12.031

[17] Ibrahim HS, Albakri ME, Mahmoud WR. Synthesis and biological evaluation of some novel thiobenzimidazole derivatives as anti-renal cancer agents through inhibition of c-MET kinase. Bioorganic Chemistry. 2019;**85**:337-348. DOI: 10.1016/j.bioorg.2019.01.006

[18] Min R, Wu W, Wang M. Discovery of 2-(1-(3-(4-Chloroxyphenyl)-3-oxo-propyl) pyrrolidine-3-yl)-1*H*-benzo [d] imidazole-4-carboxamide: A potent poly (ADP-ribose) polymerase (PARP) inhibitor for treatment of cancer. Molecules. 2019;**24**(10):1901. DOI: 10.3390/molecules24101901

[19] Gaba M, Mohan C. Development of drugs based on imidazole and benzimidazole bioactive heterocycles: Recent advances and future directions. Medicinal Chemistry Research. 2016;**25**(2):173-210. DOI: 10.1007/s00044-015-1495-5

[20] Purushottamachar P, Ramalingam S, Najar VC. Development of benzimidazole compounds for cancer therapy. In: Chemistry and Applications of Benzimidazole and its Derivatives. 2019. pp. 1-15. DOI: 10.5772/ intechopen.86691

[21] Narasimhan B, Sharma D, Kumar P. Benzimidazole: A medicinally important heterocyclic moiety. Medicinal Chemistry Research. 2012;**21**(3):269-283. DOI: 10.1007/s00044-010-9533-9

[22] Gupta S, Basu B, Das S. Benzimidazole-based palladium-*N*-heterocyclic carbene: A useful catalyst for C-C cross-coupling reaction at ambient condition. Tetrahedron. 2013;**69**(1):122-128. DOI: 10.1016/j.tet.2012.10.055

[23] Medici S, Peana M, Crisponi G. Silver coordination compounds: A new horizon in medicine. Coordination Chemistry Reviews. 2016;**327**:349-359. DOI: 10.1016/j.ccr.2016.05.015

[24] Viegas-Junior C, Danuello A, da Silva Bolzani VDS. Molecular hybridization: A useful tool in the design of new drug prototypes. Current Medicinal Chemistry. 2007;**14**(17):1829-1852. DOI: 10.2174/092986707781058805

[25] Singh A, Singh JV, Rana A. Monocarbonyl curcumin-based molecular hybrids as potent antibacterial agents. ACS Omega. 2019;**4**(7):11673-11684. DOI: 10.1021/acsomega.9b01109

[26] Yang JL, Ma YH, Li YH. Design, synthesis, and anticancer activity of novel trimethoxyphenyl-derived chalcone-benzimidazolium salts. ACS

Omega. 2019;**4**(23):20381-20393. DOI: 10.1021/acsomega.9b03077

[27] Karatas MO, Tekin S, Alici B. Cytotoxic effects of coumarin substituted benzimidazolium salts against human prostate and ovarian cancer cells. Journal of Chemical Sciences. 2019;**69**:131. DOI: 10.1007/s12039-019-1647-0

[28] Wang XQ, Chen XB, Ye PT. Synthesis and biological evaluation of novel 3-benzylcoumarin-imidazolium salts. Bioorganic & Medicinal Chemistry Letters. 2020;**30**(4):126896. DOI: 10.1016/j.bmcl.2019.126896

[29] Deng G, Zhou B, Wang J. Synthesis and antitumor activity of novel steroidal imidazolium salt derivatives. European Journal of Medicinal Chemistry. 2019; **168**:232-252. DOI: 10.1016/j.ejmech. 2019.02.025

[30] Nirmal NP, Rajput MS, Prasad RGSV. Brazilin from *Caesalpinia sappan* heartwood and its pharmacological activities: A review. Asian Pacific Journal of Tropical Medicine. 2015;**8**(6):421-430. DOI: 10.1016/j.apjtm.2015.05.014

[31] Huang M, Duan S, Ma X. Synthesis and antitumor activity of aza-brazilan derivatives containing imidazolium salt pharmacophores. MedChemComm. 2019;**10**(6):1027-1036. DOI: 10.1039/ C9MD00112C

[32] Zhou B, Liu ZF, Deng GG. Synthesis and antitumor activity of novel *N*-substituted tetrahydro-beta-carboline-imidazolium salt derivatives. Organic & Biomolecular Chemistry. 2016;**14**(39):9423-9430. DOI: 10.1039/ C6OB01495J

[33] Xu XL, Wang J, Yu CL. Synthesis and cytotoxic activity of novel 1-((indol-3-yl)methyl)-1H-imidazolium salts. Bioorganic & Medicinal Chemistry Letters. 2014;**24**(21):4926-4930. DOI: 10.1016/j.bmcl.2014.09.045

[34] Li YH, Zhou B, Shi YM. Novel 3-substituted *N*-methylcarbazole-imidazolium salt derivatives: Synthesis and cytotoxic activity. Chemical Biology & Drug Design. 2018;**92**(1):1206-1213. DOI: 10.1111/cbdd.13178

[35] Wang XQ, Ye PT, Bai MJ. Synthesis and biological activity of new bisbenzofuran-imidazolium salts. Bioorganic & Medicinal Chemistry Letters. 2020;**30**(13):127210. DOI: 10.1016/j.bmcl.2020.127210

[36] Zhang CB, Liu Y, Liu ZF. Synthesis and cytotoxic activity of novel tetrahydrobenzodifuran-imidazolium salt derivatives. Bioorganic & Medicinal Chemistry Letters. 2017;**27**(8):1808-1814. DOI: 10.1016/j.bmcl.2017.02.053

[37] Zhou Y, Duan K, Zhu L. Synthesis and cytotoxic activity of novel hexahydropyrrolo[2,3-b]indole imidazolium salts. Bioorganic & Medicinal Chemistry Letters. 2016; **26**(2):460-465. DOI: 10.1016/j. bmcl.2015.11.092

[38] Akkoc S, Kayser V, Ilhan IO. New compounds based on a benzimidazole nucleus: Synthesis, characterization and cytotoxic activity against breast and colon cancer cell lines. Journal of Organometallic Chemistry. 2017;**839**: 98-107. DOI: 10.1016/j.jorganchem. 2017.03.037

[39] Akkoc S, Kayser V, Ilhan IO. Synthesis and *in vitro* anticancer evaluation of some benzimidazolium salts. Journal of Heterocyclic Chemistry. 2019;**56**(10):2934-2944. DOI: 10.1002/ jhet.3687.svg

[40] Akkoc S. Antiproliferative activities of 2-hydroxyethyl substituted

benzimidazolium salts and their palladium complexes against human cancerous cell lines. Synthetic Communications. 2019;**49**(21):1-12. DOI: 10.1080/00397911.2019.1650187

[41] Lin ZJ, Bies J, Johnson SS. Synthesis and anti-proliferative activity of *N*, *N′*-bis-substituted 1,2,4-triazolium salts against breast cancer and prostate cancer cell Lines. Journal of Heterocyclic Chemistry. 2018;**56**(2):533-538. DOI: 10.1002/jhet.3428

[42] Wright BD, Deblock MC, Wagers PO. Anti-tumor activity of lipophilic imidazolium salts on select NSCLC cell lines. Medicinal Chemistry Research. 2015;**24**:2838-2861. DOI: 10.1007/s00044-015-1330-z

[43] Stromyer ML, Southerland MR, Satyal U. Synthesis, characterization, and biological activity of a triphenylphosphonium-containing imidazolium salt against select bladder cancer cell lines. European Journal of Medicinal Chemistry. 2020;**185**:111832. DOI: 10.1016/j.ejmech.2019.111832

[44] Shelton KL, Debord MA, Wagers PO. Synthesis, anti-proliferative activity, SAR study, and preliminary in vivo toxicity study of substituted *N,N′*-bis(arylmethyl) benzimidazolium salts against a panel of non-small cell lung cancer cell lines. Bioorganic & Medicinal Chemistry. 2017;**25**(1):421-439. DOI: 10.1016/j.bmc.2016.11.009

[45] Bansode P, Patil P, Choudhari P. Anticancer activity and molecular docking studies of ferrocene tethered ionic liquids. Journal of Molecular Liquids. 2019;**290**:111182. DOI: 10.1016/j.molliq.2019.111182

[46] KucuKbay H, Mumcu A, Tekin S. Synthesis and evaluation of novel *N*, *N′*-disubstituted benzimidazolium bromides salts as antitumor agents. Turkish Journal of Chemistry. 2016;**40**(3):393-401. DOI: 10.3906/kim-1510-15

[47] Haque RA, Iqbal MA, Ahamed MBK. Design, synthesis and structural studies of meta-xylyl linked bis-benzimidazolium salts: Potential anticancer agents against human colon cancer. Chemistry Central Journal. 2012;**6**(1):68. DOI: 10.1186/1752-153x-6-68

[48] Huda NU, Islam S, Zia M. Anticancer, antimicrobial and antioxidant potential of sterically tuned bis-*N*-heterocyclic salts. Zeitschrift für Naturforschung - Section C Journal of Biosciences. 2018;**74**(1):17-23. DOI: 10.1515/znc-2018-0095

[49] Rosenberg B, Van Camp L, Krigas T. Inhibition of cell division in *Escherichia coli* by electrolysis products from a platinum electrode. Nature. 1965;**205**: 698-699. DOI: 10.1038/205698a0

[50] Da Silva B, Habibi M, Ognier S. Silver nanocluster catalytic microreactors for water purification. The European Physical Journal Special Topics. 2016;**225**:707-714. DOI: 10.1140/epjst/e2015-50262-6

[51] Varaprasad K, Mohan YM, Vimala K. Synthesis and characterization of hydrogel-silver nanoparticle-curcumin composites for wound dressing and antibacterial application. Journal of Applied Polymer Science. 2011;**121**(2):784-796. DOI: 10.1002/app.33508

[52] Iqbal MA, Umar MI, Haque RA. Macrophage and colon tumor cells as targets for a binuclear silver(I) *N*-heterocyclic carbene complex, an anti-inflammatory and apoptosis mediator. Journal of Inorganic

Biochemistry. 2015;**146**:1-13. DOI: 10.1016/j.jinorgbio.2015.02.001

[53] Ghdhayeb MZ, Haque RA, Budagumpi S. Mono- and bis-*N*-heterocyclic carbene silver(I) and palladium(II) complexes: Synthesis, characterization, crystal structure and *in vitro* anticancer studies. Polyhedron. 2017;**121**:222-230. DOI: 10.1016/j.poly.2016.09.065

[54] Sharhan O, Heidelberg T, Hashim NM. Benzimidazolium-acridine-based silver *N*-heterocyclic carbene complexes as potential anti-bacterial and anti-cancer drug. Inorganica Chimica Acta. 2020;**504**: 119462. DOI: 10.1016/j.ica.2020.119462

[55] Habib A, Nazari VM, Iqbal MA. Unsymmetrically substituted benzimidazolium based Silver(I)-*N*-heterocyclic carbene complexes: Synthesis, characterization and in vitro anticancer study against human breast cancer and colon cancer. Journal of Saudi Chemical Society. 2019;**23**(7):795-808. DOI: 10.1016/j.jscs.2019.03.002

[56] Atif M, Bhatti HN, Haque RA. Synthesis, structure and anticancer activity of symmetrical and non-symmetrical silver(I)-*N*-heterocyclic carbene complexes. Applied Biochemistry and Biotechnology. 2020;**191**:1171-1189. DOI: 10.1007/s12010-019-03186-9

[57] Achar G, Shahini CR, Patil SA. Sterically modulated silver(I) complexes of coumarin substituted benzimidazol-2-ylidenes: Synthesis, crystal structures and evaluation of their antimicrobial and antilung cancer potentials. Journal of Inorganic Biochemistry. 2018;**183**:43-57. DOI: 10.1016/j.jinorgbio.2018.02.012

[58] Yasar S, Koprulu TK, Tekin S. Synthesis, characterisation and cytotoxic properties of *N* -heterocyclic carbene silver(I) complexes. Inorganica Chimica Acta. 2018;**479**:17-23. DOI: 10.1016/j.ica.2018.04.035

[59] Karlik O, Balcıoglu S, Karatas MO. Synthesis, structural characterization and cytotoxicity studies of T-shaped silver(I) complexes derived from 1-benzyl-3*H*-benzimidazolium *p*-toluenesulfonates. Polyhedron. 2018;**142**:63-70. DOI: 10.1016/j.poly.2017.12.033

[60] Fatima T, Haque RA, Razali MR. Effect of lipophilicity of wingtip groups on the anticancer potential of mono *N*-heterocyclic carbene silver(I) complexes: Synthesis, crystal structures and *in vitro* anticancer study. Applied Organometallic Chemistry. 2017;**31**(10): e3735. DOI: 10.1002/aoc.3735

[61] Haque RA, Choo SY, Budagumpi S. Synthesis, crystal structures, characterization and biological studies of nitrile-functionalized silver(I) *N*-heterocyclic carbene complexes. Inorganica Chimica Acta. 2015;**433**: 35-44. DOI: 10.1016/j.ica.2015.04.023

[62] Haque RA, Choo SY, Budagumpi S. Silver(I) complexes of mono- and bidentate *N*-heterocyclic carbene ligands: synthesis, crystal structures, and *in vitro* antibacterial and anticancer studies. European Journal of Medicinal Chemistry. 2015;**90**:82-92. DOI: 10.1016/j.ejmech.2014.11.005

[63] Hussaini SY, Haque RA, Asekunowo PO. Synthesis, characterization and anti-proliferative activity of propylene linked bis-benzimidazolium salts and their respective dinuclear Silver(I)-*N*-heterocyclic carbene complexes. Journal of Organometallic Chemistry. 2017;**840**:56-62. DOI: 10.1016/j.jorganchem.2017.04.011

[64] Hussaini SY, Haque RA, Fatima T. Nitrile functionalized silver(I) *N*-heterocyclic carbene complexes: DFT calculations and antitumor studies. Transition Metal Chemistry. 2018;**43**: 301-312. DOI: 10.1007/s11243-018-0216-6

[65] Akkoc S, Ozer Ilhan I, Gok Y. *In vitro* cytotoxic activities of new silver and PEPPSI palladium *N*-heterocyclic carbene complexes derived from benzimidazolium salts. Inorganica Chimica Acta. 2016;**449**:75-81. DOI: 10.1016/j.ica.2016.05.001

[66] Sahin N, Sahin-Bolukbasi S, Tahir MN. Synthesis, characterization and anticancer activity of allyl substituted *N*-Heterocyclic carbene silver (I) complexes. Journal of Molecular Structure. 2019;**1179**:92-99. DOI: 10.1016/j.molstruc.2018.10.094

[67] Ozdemir I, Ciftci O, Evren E. Synthesis, characterization and antitumor properties of novel silver(I) and gold(I) *N*-heterocyclic carbene complexes. Inorganica Chimica Acta. 2020;**506**:119530. DOI: 10.1016/j.ica.2020.119530

[68] Cevik-Yildiz E, Sahin N, Sahin-Bolukbasi S. Synthesis, characterization, and investigation of antiproliferative activity of novel Ag (I)-*N*-heterocyclic Carbene (NHC) compounds. Journal of Molecular Structure. 2020;**1199**:126987. DOI: 10.1016/j.molstruc.2019.126987

[69] Aktas A, Kelestemur U, Gok Y. 2-Morpholinoethyl-substituted *N*-heterocyclic carbene (NHC) precursors and their silver (I) NHC complexes: Synthesis, crystal structure and *in vitro* anticancer properties. Journal of the Iranian Chemical Society. 2018;**15**:131-139. DOI: 10.1007/s13738-017-1216-8

[70] Kızrak U, Ciftci O, Ozdemir I. Amine-fnctionalized silver and gold *N*-heterocyclic carbene complexes: Synthesis, characterization and antitumor properties. Journal of Organometallic Chemistry. 2019;**882**: 26-32. DOI: 10.1016/j.jorganchem.2018.12.018

[71] Sahin-Bolukbasi S, Sahin N. Novel Silver-NHC complexes: Synthesis and anticancer properties. Journal of Organometallic Chemistry. 2019;**891**: 78-84. DOI: 10.1016/j.jorganchem.2019.04.018

[72] Sahin-Bolukbasi S, Sahin N, Tahir MN. Novel *N*-heterocyclic carbene silver(I) complexes: Synthesis, structural characterization, and anticancer activity. Inorganica Chimica Acta. 2019;**486**:711-718. DOI: 10.1016/j.ica.2018.11.044

[73] Roman M, Jitaru P, Barbante C. Selenium biochemistry and its role for human health. Metallomics. 2014;**6**(1): 25-54. DOI: 10.1039/c3mt00185g

[74] Eiche E, Nothstein AK, Gottlicher J. The behaviour of irrigation induced Se in the groundwater-soil-plant system in Punjab, India. Environmental Science. Processes & Impacts. 2019;**21**(6):957-969. DOI: 10.1039/C9EM00009G

[75] Kamal A, Nazari M, Yaseen M. Green synthesis of Selenium-*N*-heterocyclic carbene compounds: Evaluation of antimicrobial and anticancer potential. Bioorganic Chemistry. 2019;**90**:103042. DOI: 103042, 10.1016/j.bioorg.2019.103042

[76] Iqbal MA, Haque RA, Ng WC. Green synthesis of mono- and di-Selenium-*N*-heterocyclic carbene adducts: Characterizations, crystal structures and pro-apoptotic activities against human colorectal cancer. Journal of

Organometallic Chemistry. 2016;**801**: 130-138. DOI: 10.1016/j.jorganchem.2015.10.023

[77] Akkoc M, Balciolu S, Gurses C. Protonated water-soluble *N*-heterocyclic carbene ruthenium(II) complexes: Synthesis, cytotoxic and DNA binding properties and molecular docking study. Journal of Organometallic Chemistry. 2018;**869**:67-74. DOI: 10.1016/j.jorganchem.2018.06.003

[78] Onar G, Gurses C, Karatas MO. Palladium (II) and ruthenium (II) complexes of benzotriazole functionalized *N*-heterocyclic carbenes: Cytotoxicity, antimicrobial, and DNA interaction studies. Journal of Organometallic Chemistry. 2019;**886**: 48-56. DOI: 10.1016/j.jorganchem.2019.02.013

[79] Lam NYS, Truong D, Burmeister H. From catalysis to cancer: Toward structure-activity relationships for benzimidazol-2-ylidene-derived *N*-heterocyclic-carbene complexes as anticancer Agents. Inorganic Chemistry. 2018;**57**(22):14427-14434. DOI: 10.1021/acs.inorgchem.8b02634

[80] Truong D, Sullivan MP, Tong KKH. Potent inhibition of thioredoxin reductase by the Rh derivatives of anticancer M(arene/Cp*) (NHC)Cl_2 complexes. Inorganic Chemistry. 2020;**59**(3):3281-3289. DOI: 10.1021/acs.inorgchem.9b03640

[81] Zhao X, Shi L, He W. Synthesis of novel NHC–Rh complexes with anti-tumor activity against MCF-7 human breast cancer cells. ARKIVOC. 2020;**vi**:94-104. DOI: 10.24820/ark.5550190.p011.129

[82] Sanchez-Mora A, Valdes H, Ramirez-Apan MT. NHC-Ir (I) complexes derived from 5, 6-dinitrobenzimidazole. Synthesis, characterization and preliminary evaluation of their *in vitro* anticancer activity. Inorganica Chimica Acta. 2019;**496**:119061. DOI: 10.1016/j.ica.2019.119061

[83] Choo KB, Mah WL, Lee SM. Palladium complexes of bidentate pyridine *N*-heterocyclic carbenes: Optical resolution, antimicrobial and cytotoxicity studies. Applied Organometallic Chemistry. 2018;**32**: e4377. DOI: 10.1002/aoc.4377

[84] Sivaram H, Tan J, Huynh HV. Syntheses, characterizations, and a preliminary comparative cytotoxicity study of gold(I) and gold(III) complexes bearing benzimidazole- and pyrazole-derived *N*-heterocyclic carbenes. Organometallics. 2012;**31**(16):5875-5883. DOI: 10.1021/om300444c

[85] Rehm T, Rothemund M, Bar A. *N,N*-dialkylbenzimidazol-2-ylidene platinum complexes - effects of alkyl residues and ancillary cis-ligands on anticancer activity. Dalton Transactions. 2018;**47**(48):17367-17381. DOI: 10.1039/C8DT03360A

Section 2

Various Biological Properties

Chapter 3

Benzimidazole: Pharmacological Profile

Mahender Thatikayala, Anil Kumar Garige and Hemalatha Gadegoni

Abstract

Benzimidazole is a bicyclic heterocyclic aromatic compound in which benzene fused to imidazole moiety. Benzimidazole holds a vital role in the field of medicinal chemistry which possesses wide variety of pharmacological activities like antibacterial, anti cancer, antifungal, antileishmanial, anti tubercular, anti viral and anti malarial respectively, hence the benzimidazole moiety attracting the medicinal chemist to synthesize the different benzimidazole derivatives with wide variety of pharmacological activities. The book chapter mainly discussed the anti cancer, anti HIV, antileishmanial and anti tubercular activites of recently synthesized benzimidazole derivatives.

Keywords: benzimidazole, anti cancer, anti HIV, antileishmanial, anti tubercular

1. Introduction

Benzimidazole is bicyclic heterocyclic aromatic compound in which benzene ring fused to 4 and 5 position of imidazole ring, it contain two nitrogen atoms at 1 and 3 position exhibit both acidic and basic nature called amphotericin nature and exists in two equivalent tautomeric forms, when the hydrogen present at first position nitrogen atom possess acidic nature, when the hydrogen present at third position nitrogen atom possess basic nature (**Figures 1**) [1]. Benzimidazole is a very important important pharmacophore among all the heterocyclic compounds due to its important pharmacological activities like anti-Alzheimer [2], antibacterial [3], anti cancer [4], antidiabetic [5], antifungal [6], anti HIV [7], anti leishmanial [8], anti inflammatory [9], analgesic [9], anti malarial [10], anti microbial [11] and anti tubercular [12] activity, there are many benzimidazole derivatives are using to treat many diseases, few presently marketing drugs contain benzimidazole moiety are the bezitramide using as an analgesic, ridinilazole sing as antibacterial, the candesartan, mibefradil using as antihypertensive drugs, mebendazole, albendazole, thiabendazole, and flubendazole usng as antihelminthics, astemizole, bilastine using as antihistamines, pantoprazole, lansoprazole, esomeprazole, ilaprazole using as proton pump inhibitors, bendamustine, selumetinib, galeterone, pracinostat using as antitumor agents and enviradine, samatasvir, and maribavir using as antiviral agents (**Figures 2**) [13–17].

IntechOpen

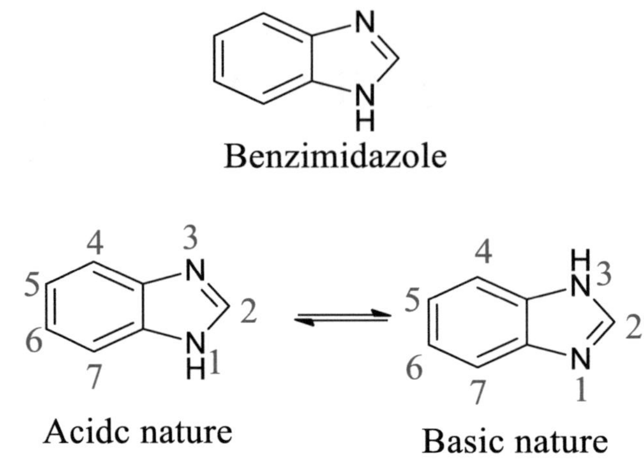

Figure 1.
Chemistry of benzimidazole [1].

2. Pharmacological profile of benzimidazole derivatives

2.1 Anti cancer activity

In the year of 2019 Tahlan et al., reported the synthesis and anti cancer activity of the new benzimidazole derivatives, among all the derivatives the compound **1** (**Figure 3**) found to be best activity at IC_{50} value of 4.53 μM against the human colorectal cancer cell line [4], same authors in 2018 reported the compound **2** (**Figure 3**) showed best activity at IC_{50} value of 4.12 μM against the human colorectal carcinoma cell line (HCT116) [18], same year few authors reported the synthesis, anti anti cancer activity of the new benzimidazole derivatives, Aikman et al., reported the compound **3** (**Figure 3**) found to be best active compound at EC50 value of 5 ± 2 μM against the melanoma (A375) cells [19], Mohamed et al., reported the compound **4** (**Figure 3**) showed best activity at IC_{50} value of 80,35, 72 μg/ml against the against human breast adenocarcinoma (MCF-7), human lung carcinoma (A549), human epitheloid cervix carcinoma (HELA) [20], Gohary et al., reported the compound **5** (**Figure 3**) showed significant activity at IC_{50}value of 0.022, 0.014, 0.015 μM against the against liver cancer (HepG2), colon cancer (HCT-116), breast cancer (MCF-7) cells [21], in 2017 Wang et al., reported the synthesis, anti-cancer activity of the chrysin benzimidazole derivatives, the compound **6** (**Figure 3**) showed significant activity at IC_{50} values of 25.72 ± 3.95 μM against MFC cells [22] and Yadav et al., reported the anti cancer activity of synthesized the 2-(*1H*-benzo[d]imidazol-2-ylthio)acetami do)-*N*-(substituted-4-oxothiazolidin-3-yl)acetamides, the compound **7, 8** (**Figure 3**) showed significant activity at IC_{50} value of 0.00005, 0.00012 μM/ml against HCT116 cell line [23], Onnis et al., reported the anti cancer activity of benzimidazolehydrazones, the compound **9** (**Figure 3**) showed excellent activity at IC_{50} value of 0.98 ± 0.02 μM against human T-lymphoblastic leukemia (CEM) cells [24].

In 2015 few authors worked on synthesis of benzimidazole and evaluated the nti-cancer activit, the Gao et al., reported the compound **10** (**Figure 3**) showed good activity at IC_{50} value of 2.68 μM against K562 and HepG-2 cells [25], Kamal et al., reported the compound **11** (**Figure 3**) found to be best at IC_{50} value of 1.8 μM against

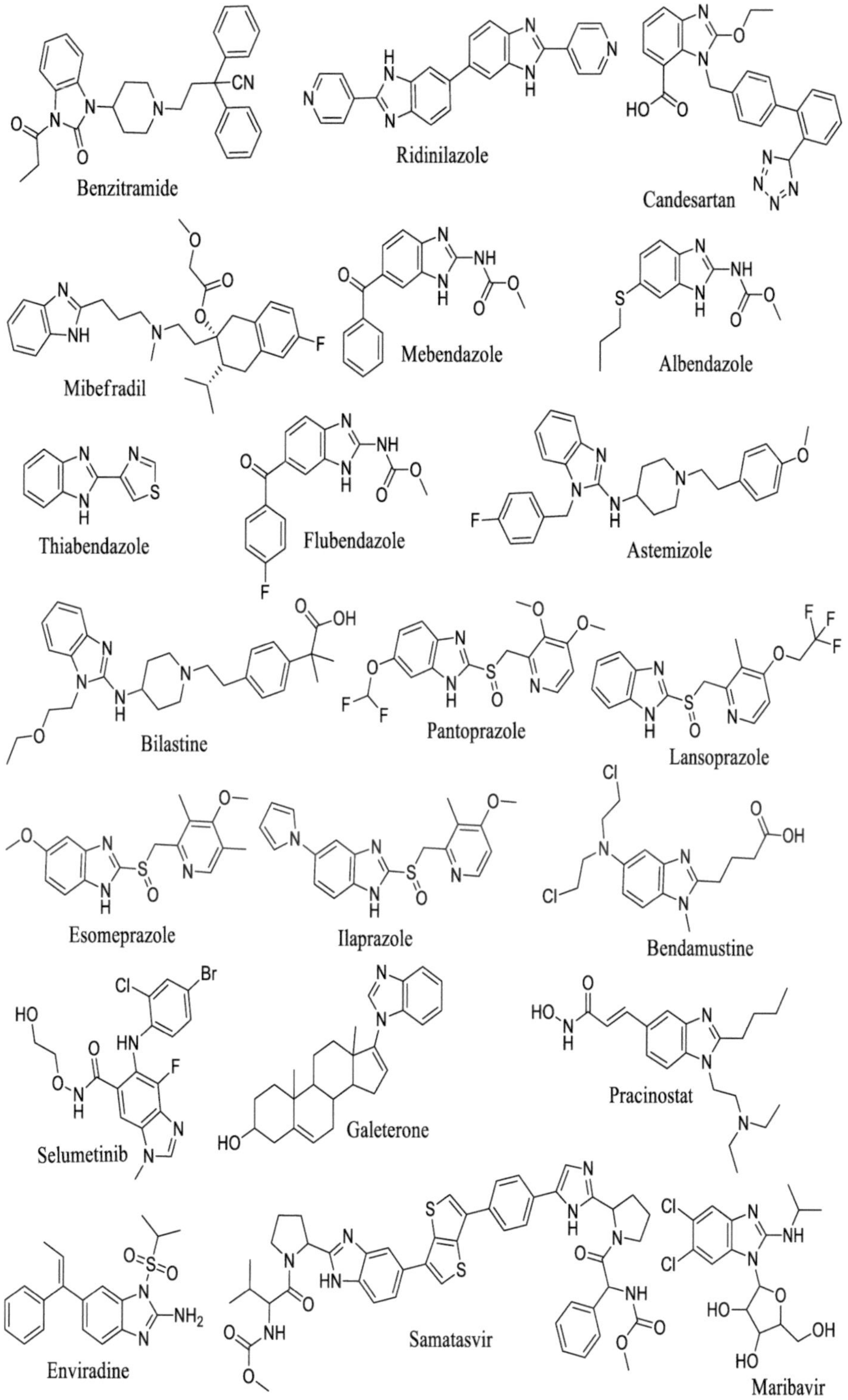

Figure 2.
Structures of marketing drugs containing benzimidazole moiety [13–17].

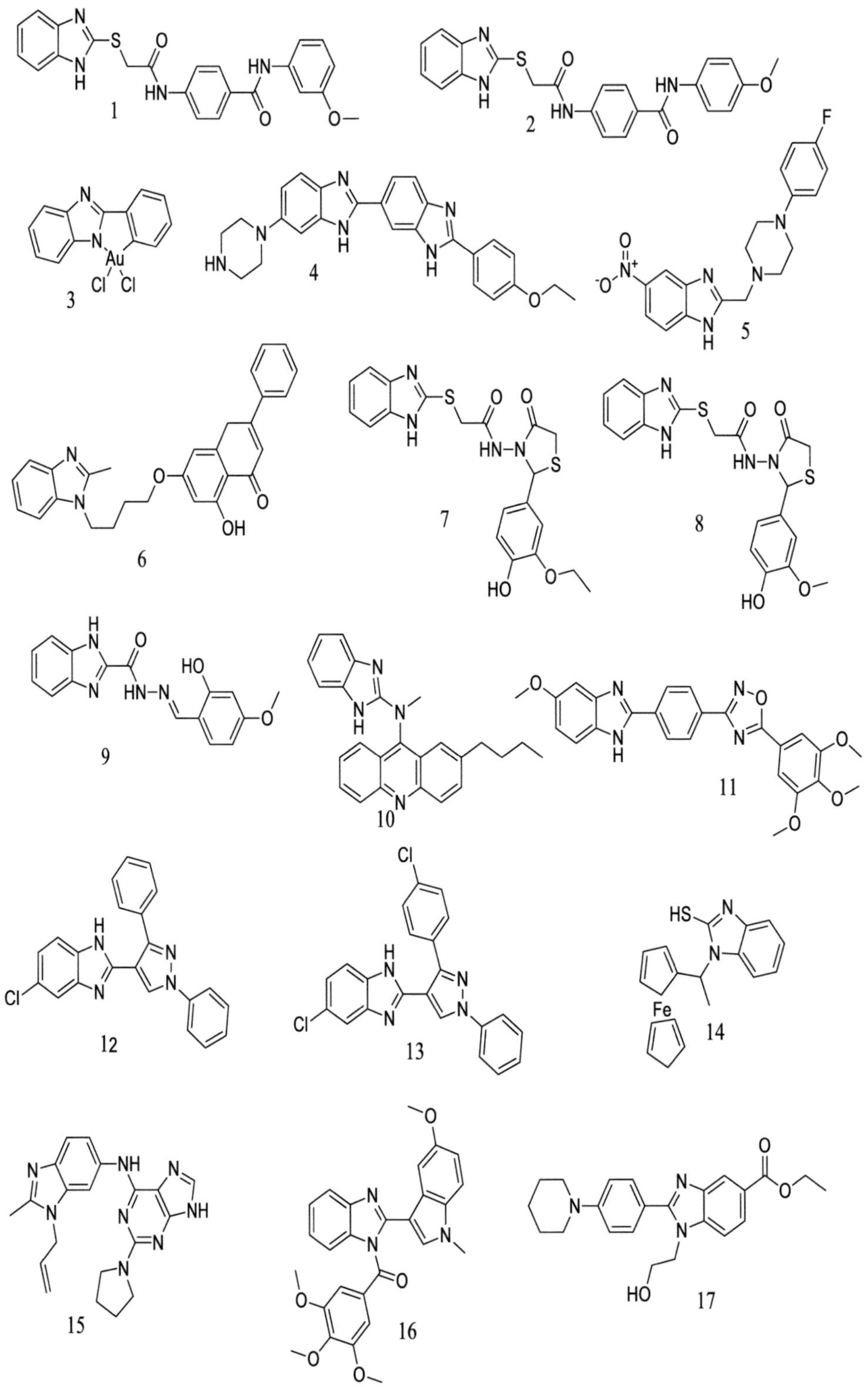

Figure 3.
Structures of effective anticancer compounds.

most of the tumor cell lines [26], T.S. Reddy et al., reported the compounds **12, 13** (**Figure 3**) showed best anti-cancer activity with IC_{50} values of 1.81, 0.83, 1.76, 1.13, 0.95, 1.57 μM against lung (A549), breast (MCF-7), cervical (HeLa) human tumor cell lines [27], Rodionov et al., reported the compound **14 (Figure 3)** found to be good activity with 87% tumor growth inhibition against carcinoma75 [28], Sharma et al., reported the Compound **15** (**Figure 3**) showed maximum activity at GI_{50} values of 3.16, 2, 1.36 μM against colon cancer, CNS cancer and ovarian cancer [29] and Wang et al., reported the compound **16** (**Figure 3**) showed excellent activity at GI_{50} values of 2.4, 3.8, 5.1 μM against human lung adenocarcinoma cells (A549), human liver hapatocellular carcinoma (HepG2), human breast carcinoma cells (MCF-7) [30].

In 2014 Yoon et al., evaluated the anti cancer activity of synthesized novel benzimidazole derivatives, the compounds **17** (**Figure 3**) **18** (**Figure 4**) found to be good at IC_{50} value of 49.63, 46.33, 62.43, 42.30 μM against breast cancer cells (MCF-7), triple-negative breast cancer cells (MDA-MB-468) [31], same year Wang et al., reported anticancer activity of benzimidazole-2-urea derivates, the compound **19** (**Figure 4**) showed significant activity at IC_{50} value range of 0.006 to1.774 μM against the K562, A431, HepG2, Hela, MDA-MB-435S cancer cells [32], Salahuddin et al., reported the compound **20** (**Figure 4**) showed best anti cancer activity at a percentage growth of 36.23, 47.56 against Breast cancer (MDA-MB-468), Melanoma (SK-MEL-28) cells [33], Paul et al., reported the compound **21** (**Figure 4**) found to be good anticancer activity at GI_{50} values of 0.34, 0.31 μM against colon cancer cell lines, prostate cancer cell lines [34], Madabushi et al., reported the compound **22** (**Figure 4**) showed best anticancer activity at IC_{50} values of 5.2, 9.8, 12.3, 11.1 μM against A549, MCF7, DU145, HeLa human cancer cell lines [35] and Guan et al., reported the compound **23** (**Figure 4**) showed significant anticancer activity with IC_{50} values of 0.098, 0.15, 0.13 μM against SGC-7901, A-549, HT-1080 human cancer cell lines [36].

In 2013 Sharma et al., reported the anti cancer activity of synthesized the benzimidazole quinazoline hybrids, the compound **24** (**Figure 4**) found to be activity with percentage growth of inhibition of 98, 94.2, 94.3, 97.5 against leukemia (K-562, SR), colon (HT29), melanoma (LOX IMVI) human cancer cell lines [37], in the same year Husain et al., reported the synthesis and the anti cancer activity of benzimidazole clubbed with triazolo-thiadiazoles and triazolo-thiadiazines, the compound **25** (**Figure 4**) found to be maximum activity with growth inhibition with GI_{50} values ranging from 0.20 to 2.58 mM against eukemia cell lines [38], Nassan et al., reported the anti cancer activity of synthesized novel 1,2,3,4 tetrahydro[1,2,4]triazino[4,5-a] benzimidazoles, the compound **26** (**Figure 4**) showed excellent activity at IC_{50} value of 0.0390 μM against human breast adenocarcinoma cell line (MCF7) [39] and Hranjec et al., reported the anti cancer activity of synthesized the novel benzimidazole schiff bases, the compound **27, 28** (**Figure 4**) found to be significant activity at IC_{50} values of 4.73, 0.96, 3.24, 1.67 μM against HeLa, WI38 cell lines [40].

2.2 Anti HIV activity

In the year of 2020 Srivastava et al., reported the synthesis and anti HIV activity of the new benzimidazole derivatives, among all the derivatives the compound **29** (**Figure 5**) found to be best activity at IC_{50} value of 0.386×10^{-5} μM against HIV-1 [7], Iannazzo et al., reported the synthesis and anti HIV activity of the new benzimidazole derivatives, among all the derivatives the compound **30** (**Figure 5**) showed best activity at IC_{50} value of 0.09 μg/mL against HIV-1 [41], Yadav et al., reported the anti HIV activity of synthesized benzimidazole derivatives, in all the synthesized

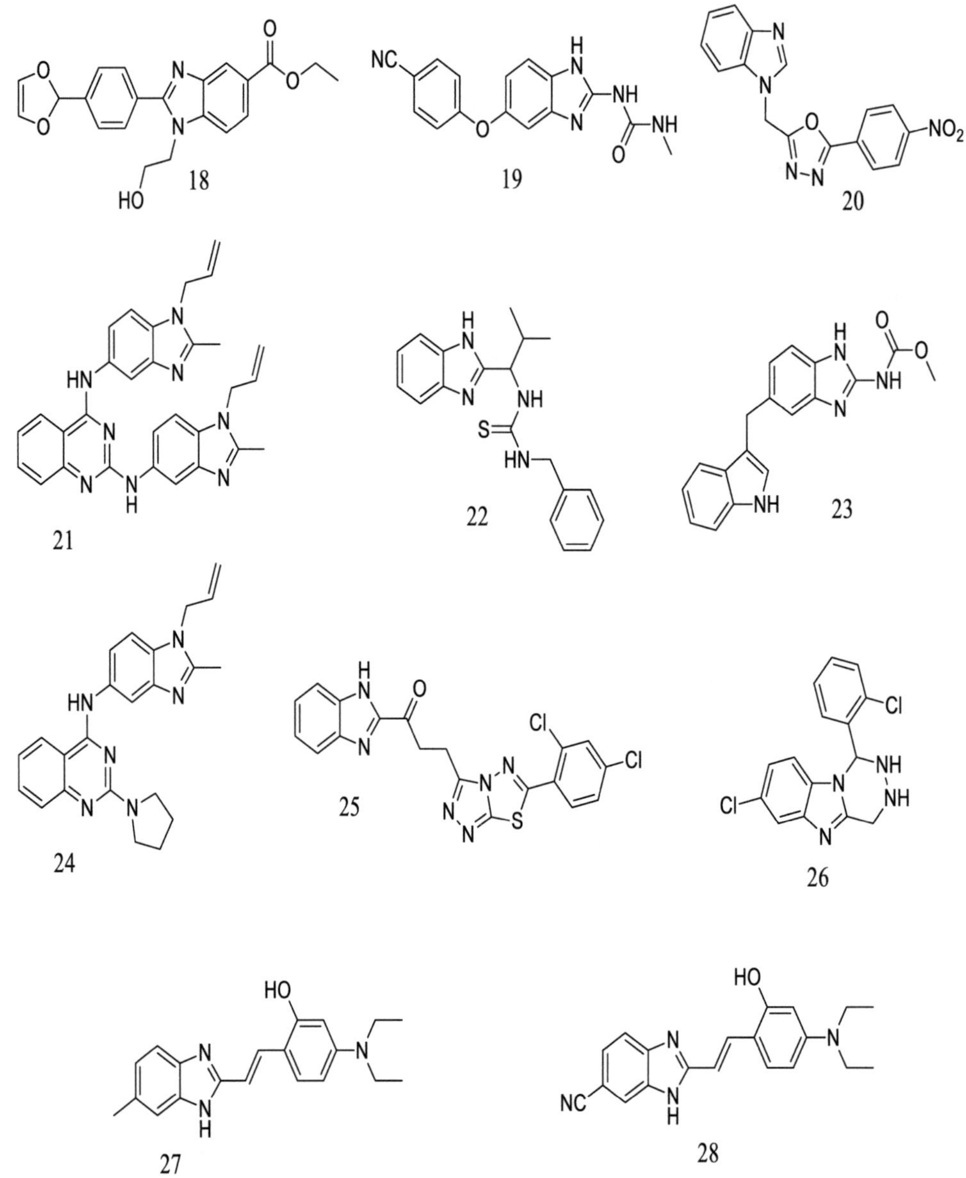

Figure 4.
Structures of effective anticancer compounds.

derivatives the compounds **31–34** (**Figure 5**) found to be best active compounds with more than 50% of RT inhibition at concentration of 20 μM against HIV-1 [42], same year Pan et al., evaluated the anti HIV activity of synthesized benzimidazoles, the compounds **35, 36** (**Figure 5**) found to be significant activity with IC_{50} values of 3.45, 58.03 nM against HIV-1 [43], Masoudi et al., synthesized the new benzimidazole derivatives, evaluated the anti HIV activity, among all the synthesized derivatives, compounds **37** (**Figure 5**) found to be significant activity at EC_{50} 1.15 μg/mL against HIV-1 and HIV-2 [44].

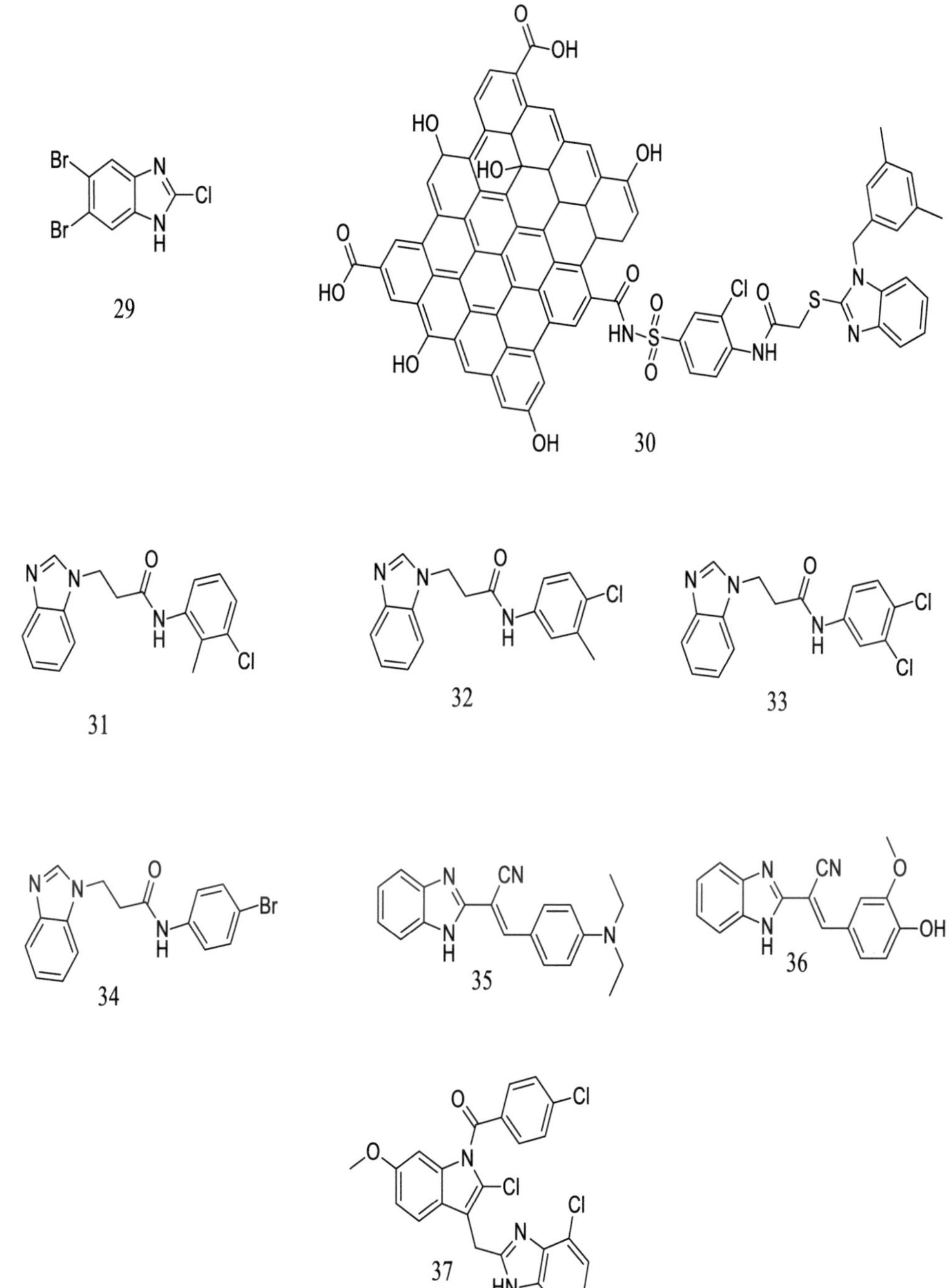

Figure 5.
Structures of effective anti-HIV compounds.

2.3 Anti leishmanial activity

M. Tonelli et al., reported the antileishmanial activity of newly synthesized benzimidazole derivatives, among all the derivatives compound **38** (**Figure 6**) found to be

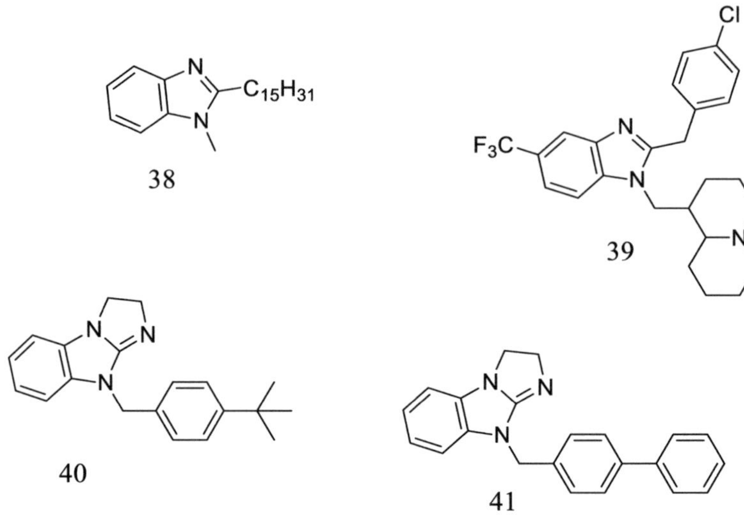

Figure 6.
Structures of effective anti-leishmanial compounds.

significant inhibition of promastigotes, amastigotes of *Leishmania tropica, Leishmania infantum* at IC_{50} values of 0.19, 0.34, 0.31 μM and compound **39** (**Figure 6**) inhibited promastigotes of *Leishmani infantum* at IC_{50} value of 3.70, 4.76 μM [8], Oh et al., reported the antileishmanial activity of newly synthesized benzimidazole derivatives, among all the derivatives compound **40, 41** (**Figure 6**) found to be most active against promastigotes, amastigotes of *Leishmania donavani* at EC_{50} values of 1.25, 3.05, 1.48 5.29 μM [45].

2.4 Anti tubercular activity

In the year of 2019 S. Manivannan et al., reported the synthesis anti tubercular activity of benzimidazole derivatives, among all the derivatives compound **42, 43** (**Figure 7**) showed best anti tubercular activity with MIC values of 6.5, 6.5, 12.5, 6.5, 12.5, 6.5 μg/mL against *Mycobaterium tuberculosis* H37Rv, drug-resistant, drug-susceptible strains [12], previous year Mohanty et al., reported the anti tubercular activity of synthesized the novel azo derivatives of benzimidazoles, in all the derivatives the compounds **44** (**Figure 7**) showed best activity at IC_{50} value of 0.119 μM/mL against *Mycobaterium tuberculosis* [46], before previous year Yadav et al., synthesized the benzimidazole derivatives, reported the anti tubercular activity the compounds **45–53** (**Figure 7**) at MIC value of 12.5 μg/mL against *Mycobaterium tuberculosis* strains of H37Rv [47]. In the year of 2015 Ramprasad et al., reported the synthesis, anti tubercular activity of the imidazo[2,1-b][1,3,4]thiadiazole-benzimidazole derivatives, the compounds **54–60** (**Figures 7** and **8**) showed best activity at MIC value of 3.125 μg/mL against *Mycobaterium tuberculosis* strains of H37Rv, Species192, Specis210 [48], same year Yoon et al., evaluated the anti tubercular activity of synthesized the new benzimidazole aminoesters, the compound **61** (**Figure 8**) showed best activity with IC_{50} value of 11.52 μM against *Mycobaterium tuberculosis* strains of H37Rv [49].

In the year of 2014 many authors reported the anti tubercular activity of synthesized the new benzimidazole derivatives, Gong et al., reported the compound **62**

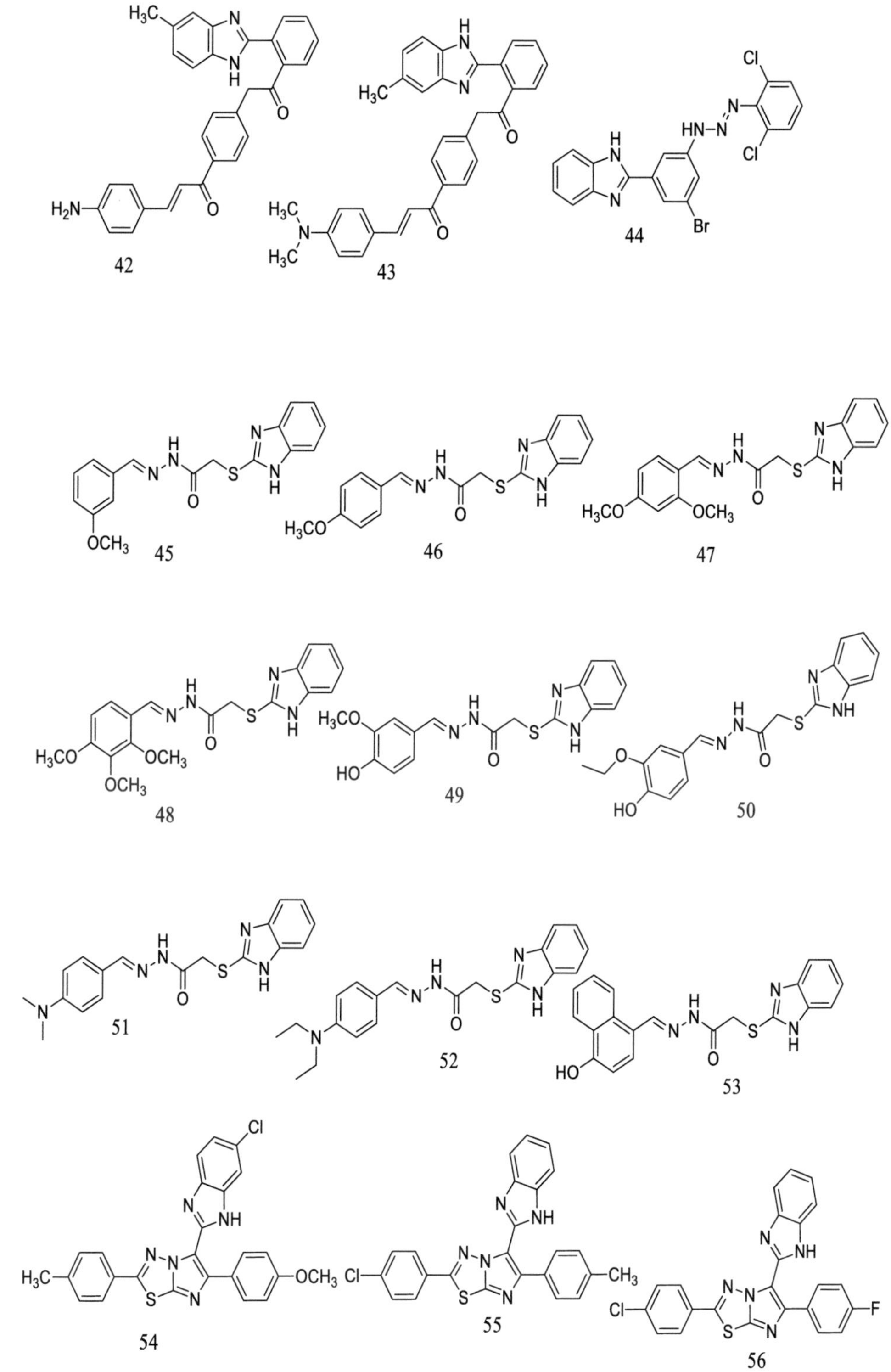

Figure 7.
Structures of effective anti-tubercular compounds.

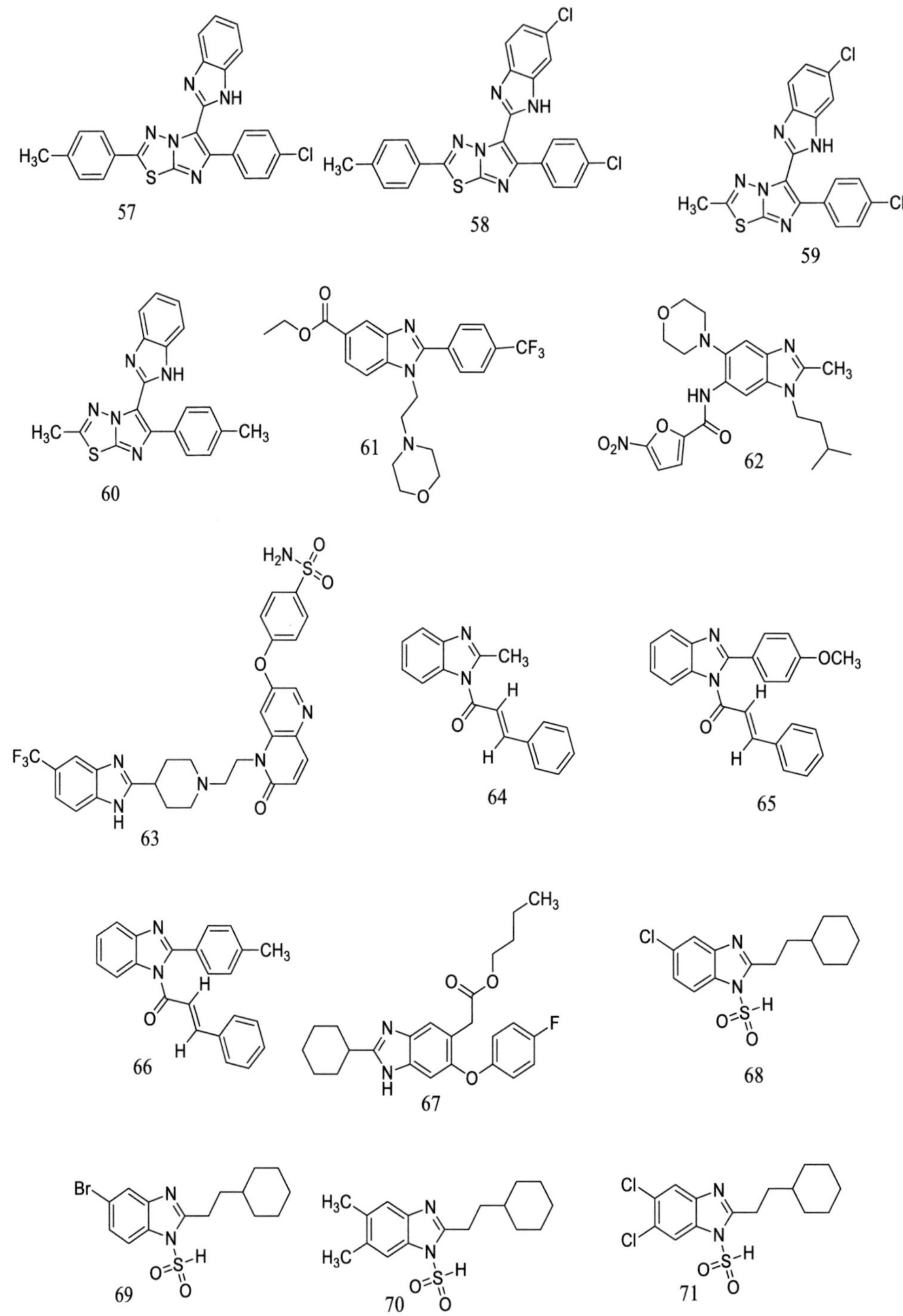

Figure 8.
Structures of effective anti-tubercular compounds.

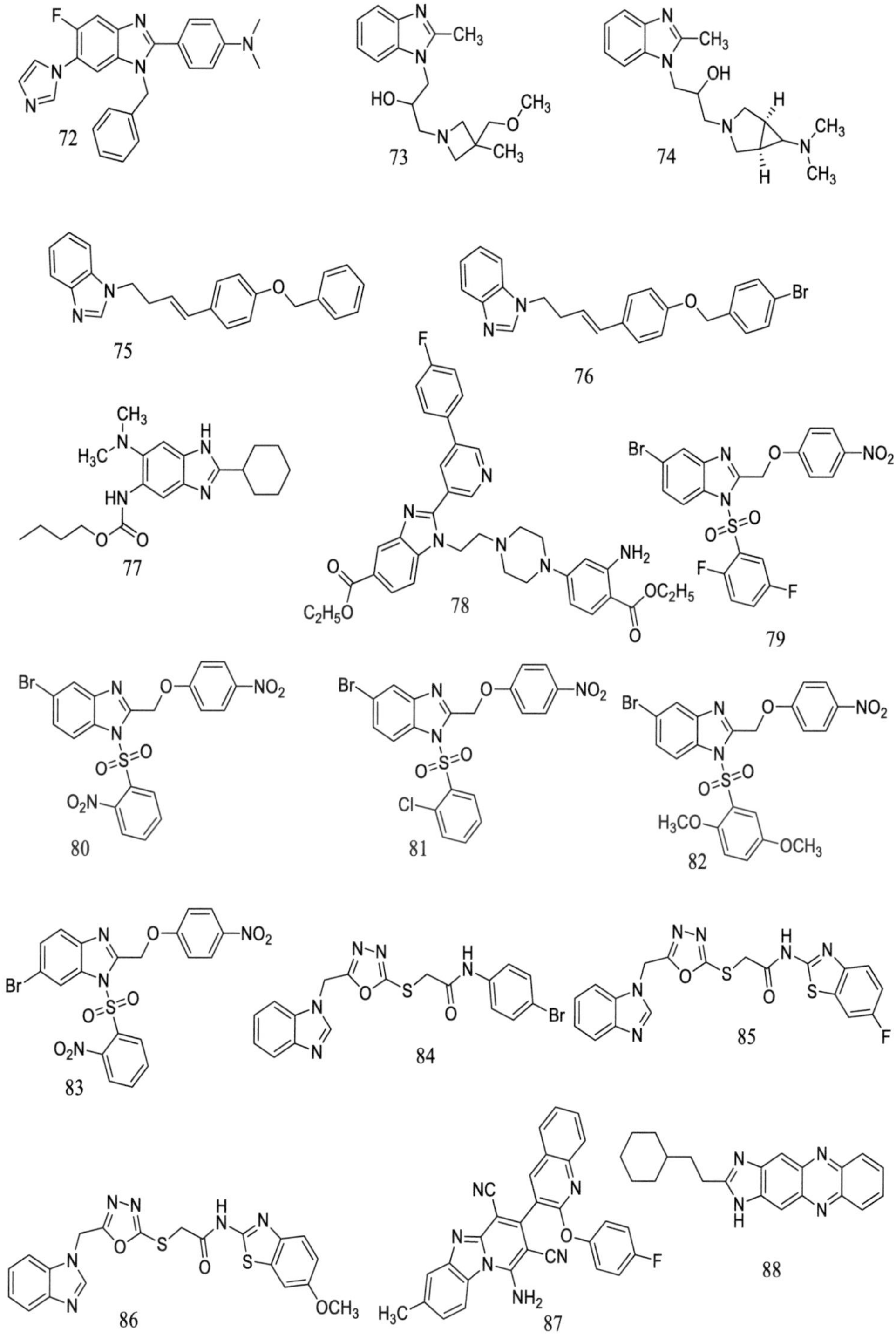

Figure 9.
Structures of effective anti-tubercular compounds.

(**Figure 8**) found to be best activity at MIC value of 0.20, 0.049 μg/mL against non-replicating *Mycobaterium tuberculosis* and replicating *Mycobacterium tuberculosis* [50], Hameed et al., reported the compound the compounds **63** (**Figure 8**) showed significant activity at MIC value of 0.19 μM against fluoroquinolone-resistant strains of *Mycobaterium tuberculosis* [51], Kalalbandi et al., reported the compounds **64–66** (**Figure 8**) showed good activity at MIC value of 3.12, 3.12, 1.6 μg/mL against *Mycobaterium tuberculosis* strains of H37Rv [52], Park et al., reported the compounds **67** (**Figure 8**) showed excellent activity at MIC value of 0.63 μg/mL against *Mycobaterium tuberculosis* strains of H37Rv [53] and Gobis et al., reported the compounds **68–71** (**Figure 8**) found to be better activity at MIC value of 0.75 μg/mL against *Mycobaterium tuberculosis* strains of H37Rv, Spec. 192, Spec. 210 [54].

In the year of 2013 also many authors evaluated the anti tubercular activity of newly synthesized benzimidazole derivatives, Nandha et al., reported the compound **72** (**Figure 9**) showed best activity at MIC value of 12.5 μg/mL against *Mycobaterium tuberculosis* strains of H37Rv [55], Birajdara et al., reported the compound **73, 74** (**Figure 9**) showed good activity at MIC value of 6.25 μg/mL against *Mycobaterium tuberculosis* strains of H37Rv [56], Anand et al., reported the compounds **75, 76** (**Figure 9**) found to be significant activity at MIC value of 1.56 μg/mL against *Mycobaterium tuberculosis* strains of H37Rv [57], Awasthi et al., reported the compound **77** (**Figure 9**) showed better activity at MIC value of 0.06 μg/mL against *Mycobaterium tuberculosis* strains of H37Rv [58], Yoon et al., reported the compound **78** (**Figure 9**) showed best activity at MIC value of 0.115, 6.12 μM against *Mycobacterium tuberculosis* H37Rv and INH-resistant *Mycobacterium tuberculosis* [59] and Ranjith et al., reported the compounds **79–83** (**Figure 9**) showed excellent activity at MIC value of 1 μg/mL against *Mycobacterium tuberculosis* H37Rv [60].

In 2012 Patel et al., reported the anti tubercular activity of synthesized the benzimidazolyl-1,3,4-oxadiazol-2ylthio-*N*-phenyl(benzothiazolyl)acetamides, among all the synthesized derivatives, the compounds **84–86** (**Figure 9**) showed best activity at MIC value of 12.5 μg/mL against *Mycobaterium tuberculosis* strains of H37Rv [61],

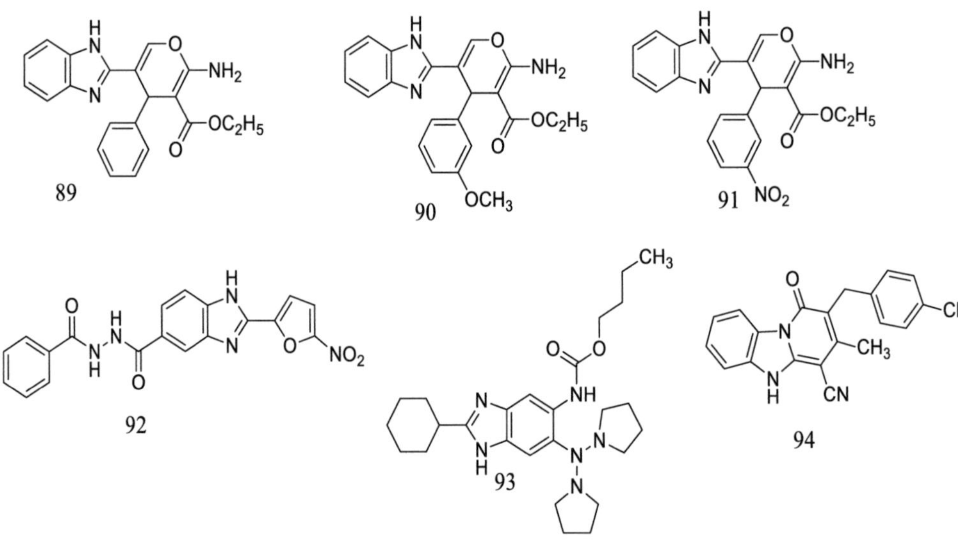

Figure 10.
Structures of effective anti-tubercular compounds.

Sangani et al., reported the synthesis and anti tubercular activity of pyrido[1,2-a] benzimidazole derivatives of beta-aryloxyquinoline, among all the derivative, the compound **87** (**Figure 9**) found to be best active compound at MIC value of 6.25 μg/mL against *Mycobaterium tuberculosis* strains of H37Rv compared with isoniazid, refampicin [62] and Gobis et al., reported the anti tubercular activity of new benzimidazoles, the compound **88** (**Figure 10**) showed best activity at MIC value of 3.1, 1.5, 3.1 μg/mL against *Mycobaterium tuberculosis* strains of H37Rv, Species 192, Species 210 [63].

In 2011 few authors reported the anti tubercular activity of synthesized benzimidaoles, Saleshier et al., reported the compounds **89–91** (**Figure 10**) found to be best activity at 10, 100mcg/ml concentrations against *Mycobaterium tuberculosis* [64], Camacho et al., reported the compound **92** (**Figure 6**) showed best activity with MIC values of 12.5 μg/mL, 6.25 μg/mL against multidrug-resistant MDR, MTB strains [65], Kumar et al., reported the compound **93** (**Figure 6**) found to be better activity at MIC99values of 1.0 μM, 1.0 μM against *Mycobaterium tuberculosis* strains of H37Rv, W210, NHN 20, NHN335, NHN382, TN587 [66] and Pieroni et al., reported the compound **94** (**Figure 10**) showed excellent activity at MIC values of 0.5 μg/mL, 1.0 μg/mL, 8.0 μg/mL against *Mycobaterium tuberculosis* strains of H37Rv [67].

3. Conclusions

The benimidazole plays in important role in the field of medicinal chemistry, many of the marketing drugs contain benzimidazole moiety are using to illness. In recent medicinal chemistry research the benzimidazole derivatives are in continuous development with many pharmacological activities such as anti-cancer, anti-HIV, anti-leishmanial, anti-tubercular, anti-malarial, anti-inflammatory, anti-diabetic, and so on, to meet pharmacological requirement. The present literature may helpful to researcher, medicinal chemist, pharmacologist to design, to synthesize, to develop pharmacologically active benzimidazole derivatives with low toxicity in future.

Acknowledgements

All the authors are thankful to managements of respective colleges for providing facilities to carry out the work.

Author contribution

Mahender Thatikayala contributed the chemistry, anti cancer, anti leishmanial and anti tubercular activity of benzimidazoles. Anil Kumar Garige, Hemalatha Gadegoni contributed the chemistry and anti HIV activity of benzimidazoles.

Funding

No funding applicable.

Conflict of interest

The authors declare no conflict of interest.

Author details

Mahender Thatikayala[1*], Anil Kumar Garige[2] and Hemalatha Gadegoni[3]

1 Avanthi Institute of Pharmaceutical Sciences, Hyderabad, Telangana, India

2 Jayamukhi Institute of Pharmaceutical Sciences, Warangal, Telangana, India

3 Pathfinder Institute of Pharmacy Education and Research, Warangal, Telangana, India

*Address all correspondence to: mahenderpharma395@gmail.com

References

[1] Singh VK, Parle A. The intriguing benzimidazole: A review. International Journal of Pharmaceutical Sciences and Research. 2019;**10**(4):1540-1552. DOI: 10.13040/IJPSR.0975-8232.10(4).1540-52

[2] Chaves S, Hiremathad A, Tomas D, Keri RS, Piemontese L, Santos MA. Exploring the chelating capacity of 2-hydroxyphenyl-benzimidazole based hybrids with multi-target ability as anti-Alzheimer's agent. New Journal of Chemistry. 2018;**42**(20):16503-16515. DOI: 10.1039/c8nj00117k

[3] Marinescu M. Synthesis of antimicrobial benzimidazole–pyrazole compounds and their biological activities. Antibiotics. 2021;**10**(1002):1-29

[4] Tahlan S, Ramasamy K, Lim SM, Shah SAA, Mani V, Narasimhan B. 4-(2-(1*H*-Benzo[d]imidazol-2-ylthio) acetamido)-*N*-(substituted phenyl) benzamides: Design, synthesis and biological evaluation. BMC Chemistry. 2019;**13**(1):1-16. DOI: 10.1186/s13065-019-0533-7

[5] Aboul-Enein HY, Rashedy AAE. Benzimidazole derivatives as antidiabetic agents. Medicinal Chemistry. 2015;**5**(7):318-325. DOI: 10.4172/2161-0444.1000280

[6] Si W, Zhang T, Li Y, She D, Pan W, Gao Z, et al. Synthesis and biological activity of novel benzimidazole derivatives as potential antifungal agents. Journal of Pesticide Science. 2016;**41**(1):15-19. DOI: 10.1584/jpestics.D15-037

[7] Srivastava R, Gupta SK, Naaz F, SenGupta PS, Yadav M, Singh VK, et al. Alkylated benzimidazoles: Design, synthesis, docking, DFT analysis, ADMET property, molecular dynamics and activity against HIV and YFV. Computational Biology and Chemistry. 2020;**89**:1-16. DOI: 10.1016/j.compbiolchem.2020.107400

[8] Tonelli M, Gabriele E, Piazza F, Basilico N, Parapini S, Tasso B, et al. Benzimidazole derivatives endowed with potent antileishmanial activity. Journal of Enzyme Inhibition and Medicinal Chemistry. 2017;**33**(1):210-226. DOI: 10.1080/14756366.2017.1410480

[9] Gaba M, Singh S, Mohan C. Benzimidazole: An emerging scaffold for analgesic and anti-inflammatory agents. European Journal of Medicinal Chemistry. 2014;**76**:494-505. DOI: 10.1016/j.ejmech.2014.01.030

[10] Farahat AA, Ismail MA, Kumar A, Wenzler T, Brun R, Paul A, et al. Indole and benzimidazole bichalcophenes: Synthesis, DNA binding and antiparasitic activity. European Journal of Medicinal Chemistry. 2018;**143**:1590-1596. DOI: 10.1016/j.ejmech.2017.10.056

[11] Marinescu M, Tudorache GD, Marton GI, Zalaru CM, Popa M, Chifiriuc MC, et al. Density functional theory molecular modeling, chemical synthesis, and antimicrobial behavior of selected benzimidazole derivatives. Journal of Molecular Structure. 2017;**1130**:463-471. DOI: 10.1016/j.molstruc.2016.10.066

[12] Manivannan S. Substituted benzimidazoles: A novel class of anti-tubercular agents. International Journal of Pharmaceutical Chemistry and Analysis. 2019;**6**(1):10-13. DOI: 10.18231/j.ijpca.2019.003

[13] Bansal Y, Silakari O. The therapeutic journey of benzimidazoles: A review. European Journal of Medicinal Chemistry. 2012;**20**(21):6208-6236. DOI: 10.1016/j.bmc.2012.09.013

[14] Njar VC, Brodie AM. Discovery and development of galeterone (TOK-001 or VN/124-1) for the treatment of all stages of prostate cancer. Journal of Medicinal Chemistry. 2015;**58**(5):2077-2087. DOI: 10.1021/jm501239f

[15] Moniruzzaman RS, Mahmud T. Quantum chemical and pharmacokinetic studies of some proton pump inhibitor drugs. American Journal of Biomedical Sciences and Research. 2019;**2**(1):3-8. DOI: 10.34297/AJBSR.2019.02.000562

[16] Scholten WK, Christensen AE, Olesen AE, Drewes AM. Quantifying the adequacy of opioid analgesic consumption globally: An updated method and early findings. American Journal of Public Health. 2019;**109**(1): 52-57. DOI: 10.2105/AJPH.2018.304753

[17] Tahlan S, Kumar S, Ramasamy K, Lim SM, Shah SAA, Mani V, et al. Design, synthesis and biological profile of heterocyclic benzimidazole analogues as prospective antimicrobial and antiproliferative agents. BMC Chemistry. 2019;**13**(50):1-15. DOI: 10.1186/s13065-019-0567-x

[18] Tahlan S, Ramasamy K, Lim SM, Shah SAA, Mani V, Narasimhan B. Design, synthesis and therapeutic potential of 3-(2-(*1H*-benzo[d]imidazol-2-ylthio)acetamido)-*N*-(substituted phenyl)benzamide analogues. Chemistry Central Journal. 2018;**12**(1):1-12. DOI: 10.1186/s13065-018-0513-3

[19] Aikman B, Wenzel M, Mósca A, de Almeida A, Klooster W, Coles S, et al. Gold(III) pyridine-benzimidazole complexes as aquaglyceroporin inhibitors and antiproliferative agents. Inorganics. 2018;**6**(4):1-16. DOI: 10.3390/inorganics6040123

[20] Mohamed LW, Taher AT, Rady GS, Ali MM, Mahmoud AE. Synthesis and biological evaluation of certain new benzimidazole derivatives as cytotoxic agents new cytotoxic benzimidazoles. Der Pharma Chemica. 2018;**10**(5): 112-120

[21] El-Gohary NS, Shaaban MI. Synthesis and biological evaluation of a new series of benzimidazole derivatives as antimicrobial, antiquorum-sensing and antitumor agents. European Journal of Medicinal Chemistry. 2017;**131**:1-29. DOI: 10.1016/j.ejmech.2017.03.018

[22] Wang Z, Deng X, Xion S, Xiong R, Liu J, Zou L, et al. Design, synthesis and biological evaluation of chrysin benzimidazole derivatives as potential anticancer agents. Natural Product Research. 2017;**32**(24):1-10. DOI: 10.1080/14786419.2017.1389940

[23] Yadav S, Narasimhan B, Lim SM, Ramasamy K, Vasudevan M, Shah SAA, et al. Synthesis, characterization, biological evaluation and molecular docking studies of 2-(*1H*-benzo[d]imidazol-2-ylthio)-*N*-(substituted 4-oxothiazolidin-3-yl) acetamides. Chemistry Central Journal. 2017;**11**(1): 1-12. DOI: 10.1186/s13065-017-0361-6

[24] Onnis V, Demurtas M, Deplano A, Balboni G, Baldisserotto A, Manfredini S, et al. Design, synthesis and evaluation of antiproliferative activity of new benzimidazolehydrazones. Molecules. 2016;**21**(5):1-9. DOI: 10.3390/molecules21050579

[25] Gao C, Li B, Zhang B, Sun Q, Li L, Li X, et al. Synthesis and biological evaluation of benzimidazole acridine derivatives as potential DNA-binding

and apoptosis-inducing agents. Bioorganic and Medicinal Chemistry. 2015;**23**(8):1800-1807. DOI: 10.1016/j.bmc.2015.02.036

[26] Kamal A, Reddy TS, Vishnuvardhan MVPS, Nimbarte VD, Subba Rao AV, Srinivasulu V, et al. Synthesis of 2-aryl-1,2,4-oxadiazolo-benzimidazoles: Tubulin polymerization inhibitors and apoptosis inducing agents. Bioorganic and Medicinal Chemistry. 2015;**23**(15):4608-4623. DOI: 10.1016/j.bmc.2015.05.060

[27] Reddy TS, Kulhari H, Reddy VG, Bansal V, Kamal A, Shukla R. Design, synthesis and biological evaluation of 1,3-diphenyl-*1H*-pyrazole derivatives containing benzimidazole skeleton as potential anticancer and apoptosis inducing agents. European Journal of Medicinal Chemistry. 2015;**101**:790-805. DOI: 10.1016/j.ejmech.2015.07.031

[28] Rodionov AN, Zherebker KY, Snegur LV, Korlyukov AA, Arhipov DE, Peregudov AS, et al. Synthesis, structure and enantiomeric resolution of ferrocenylalkyl mercaptoazoles. Antitumor activity in vivo. Journal of Organometallic Chemistry. 2015;**783**: 83-91. DOI: 10.1016/j.jorganchem.2015.01.031

[29] Sharma A, Luxami V, Paul K. Purine-benzimidazole hybrids: Synthesis, single crystal determination and in vitro evaluation of antitumor activities. European Journal of Medicinal Chemistry. 2015;**93**:414-422. DOI: 10.1016/j.ejmech.2015.02.036

[30] Wang YT, Qin YJ, Yang N, Zhang YL, Liu CH, Zhu HL. Synthesis, biological evaluation, and molecular docking studies of novel 1-benzene acyl-2-(1-methylindol-3-yl)-benzimidazole derivatives as potential tubulin polymerization inhibitors. European Journal of Medicinal Chemistry. 2015;**99**:125-137. DOI: 10.1016/j.ejmech.2015.05.021

[31] Yoon YK, Ali MA, Wei AC, Choon TS, Osma H, Parang K, et al. Synthesis and evaluation of novel benzimidazole derivatives as sirtuin inhibitors with antitumor activities. Bioorganic & Medicinal Chemistry. 2014;**22**(2):703-710. DOI: 10.1016/j.bmc.2013.12.029

[32] Wang W, Kong D, Cheng H, Tan L, Zhang Z, Zhuang X, et al. New benzimidazole-2-urea derivates as tubulin inhibitors. Bioorganic & Medicinal Chemistry Letters. 2014; **24**(17):4250-4253. DOI: 10.1016/j.bmcl.2014.07.035

[33] Salahuddin, Shaharyar M, Mazumder A, Ahsan MJ. Synthesis, characterization and anticancer evaluation of 2-(naphthalen-1-ylmethyl/naphthalen-2-yloxymethyl)-1-[5-(substituted phenyl)-[1,3,4]oxadiazol-2-ylmethyl]-*1H*-benzimidazole. Arabian Journal of Chemistry. 2014;7(4):418-424. DOI: 10.1016/j.arabjc.2013.02.001

[34] Paul K, Sharma A, Luxami V. Synthesis and in vitro antitumor evaluation of primary amine substituted quinazoline linked benzimidazole. Bioorganic and Medicinal Chemistry Letters. 2014;**24**(2):624-629. DOI: 10.1016/j.bmcl.2013.12.005

[35] Madabhushi S, Mallu KKR, Vangipuram VS, Kurva S, Poornachandra Y, Ganesh KC. Synthesis of novel benzimidazole functionalized chiral thioureas and evaluation of their antibacterial and anticancer activities. Bioorganic and Medicinal Chemistry Letters. 2014;**24**(20):4822-4825. DOI: 10.1016/j.bmcl.2014.08.064

[36] Guan Q, Han C, Zuo D, Zhai M, Li Z, Zhang Q, et al. Synthesis and

evaluation of benzimidazole carbamates bearing indole moieties for antiproliferative and antitubulin activities. European Journal of Medicinal Chemistry. 2014;**87**:306-315. DOI: 10.1016/j.ejmech.2014.09.071

[37] Sharma A, Luxami V, Paul K. Synthesis, single crystal and antitumor activities of benzi midazole quinazoline hybrids. Bioorganic and Medicinal Chemistry Letters. 2013;**23**(11): 3288-3294. DOI: 10.1016/j.bmcl.2013.03.107

[38] Husain A, Rashid M, Shaharyar M, Siddiqui AA, Mishra R. Benzimidazole clubbed with triazolo-thiadiazoles and triazolo-thiadiazines:New anticancer agents. European Journal of Medicinal Chemistry. 2013;**62**:785-798. DOI: 10.1016/j.ejmech.2012.07.011

[39] El-Nassan HB. Synthesis, antitumor activity and SAR study of novel [1,2,4] triazino[4,5-a]benzimidazole derivatives. European Journal of Medicinal Chemistry. 2012;**53**:22-27. DOI: 10.1016/j.ejmech.2012.03.028

[40] Hranjec M, Starcevic K, Pavelic SK, Lucin P, Pavelic K, Karminski ZG. Synthesis, spectroscopic characterization and antiproliferative evaluation in vitro of novel Schiff bases related to benzimidazoles. European Journal of Medicinal Chemistry. 2011;**46**(6): 2274-2279. DOI: 10.1016/j.ejmech.2011.03.00

[41] Iannazzo D, Pistone A, Ferro S, DeLuca L, Monforte AM, Romeo R, et al. Graphene quantum dots based systems as HIV inhibitors. Bioconjugate Chemistry. 2018;**29**(9):1-38. DOI: 10.1021/acs.bioconjchem.8b00448

[42] Yadav G, Ganguly S, Murugesan S, Dev A. Synthesis, anti-HIV, antimicrobial evaluation and structure activity relationship studies of some novel benzimidazole derivatives. Anti Infective Agents. 2015; **13**(1):65-77

[43] Pan T, He X, Chen B, Chen H, Geng G, Luo H, et al. Development of benzimidazole derivatives to inhibit HIV-1 replication through protecting APOBEC3G protein. European Journal of Medicinal Chemistry. 2015;**95**:500-513. DOI: 10.1016/j.ejmech.2015.03.050

[44] Al-Masoudi NA, Jafar NNA, Abbas LJ, Baqir SJ, Pannecouque C. Synthesis and anti-HIV activity of new benzimidazole, benzothiazole and carbohyrazide derivatives of the anti-inflammatory drug indomethacin. Zeitschrift für Naturforschung. 2011;**66**: 953-960

[45] Oh S, Kim S, Kong S, Yang G, Lee N, Han D, et al. Synthesis and biological evaluation of 2,3-dihydroimidazo[1,2-a] benzimidazole derivatives against *Leishmania donovani* and *Trypanosoma cruzi*. European Journal of Medicinal Chemistry. 2014;**84**:395-403. DOI: 10.1016/j.ejmech.2014.07.038

[46] Mohanty SK, Khuntia A, Yellasubbaiah N, Ayyanna C, Naga Sudha B, Harika MS. Design, synthesis of novel azo derivatives of benzimidazole as potent antibacterial and anti tubercular agents. Beni-Suef University Journal of Basic and Applied Sciences. 2018;**7**(4):1-6. DOI: 10.1016/j.bjbas.2018.07.009

[47] Yadav S, Narasimhan B, Lim SM, Ramasamy K, Vasudevan M, SAA S, et al. Synthesis and evaluation of antimicrobial, antitubercular and anticancer activities of benzimidazole derivatives. Egyptian Journal of Basic and Applied Sciences. 2018;**5**(1): 100-109. DOI: 10.1016/j.ejbas.2017.11.001

[48] Ramprasad J, Nayak N, Dalimba U, Yogeeswari P, Sriram D, Peethambar SK, et al. Synthesis and biological evaluation of new imidazo [2,1-b][1,3,4] thiadiazole-benzimidazole derivatives. European Journal of Medicinal Chemistry. 2015;**95**:49-63. DOI: 10.1016/j.ejmech.2013.06.025

[49] Ramprasad J, Nayak N, Dalimba U, Yogeeswari P, Sriram D, Peethambar SK, et al. Synthesis and biological evaluation of new imidazo[2,1-b][1,3,4]thiadiazole-benzimidazole derivatives. European Journal of Medicinal Chemistry. 2015;**95**: 49-63. DOI: 10.1016/j.ejmech.2015.03.024

[50] Gong Y, Karakaya SS, Guo X, Zheng P, Gold B, Ma Y, et al. Benzimidazole-based compounds kill Mycobacterium tuberculosis. European Journal of Medicinal Chemistry. 2014;**75**: 336-353. DOI: 10.1016/j.ejmech.2014.01.039

[51] Hameed PS, Raichurkar A, Madhavapeddi P, Menasinakai S, Sharma S, Kaur P, et al. Benzimidazoles: Novel mycobacterial gyrase inhibitors from scaffold morphing. ACS Medicinal Chemistry Letters. 2014;**5**(7):820-825. DOI: 10.1021/ml5001728

[52] Kalalbandi VKA, Seetharamappa J, Katrahalli U, Bhat KG. Synthesis, crystal studies, anti-tuberculosis and cytotoxic studies of 1-[(2E)-3-phenylprop-2-enoyl]-*1H*-benzimidazole derivatives. European Journal of Medicinal Chemistry. 2014;**79**:194-202. DOI: 10.1016/j.ejmech.2014.04.017

[53] Park B, Awasthi D, Chowdhury SR, Melief EH, Kumar K, Knudson SE, et al. Design, synthesis and evaluation of novel 2,5,6-trisubstituted benzimidazoles targeting FtsZ as antitubercular agents. Bioorganic and Medicinal Chemistry. 2014;**22**(9): 2602-2612. DOI: 10.1016/j.bmc.2014.03.035

[54] Gobis K, Foks H, Serocki M, Augustynowicz-Kopeć E, Napiórkowska A. Synthesis and evaluation of in vitro antimycobacterial activity of novel *1H*-benzo[d]imidazole derivatives and analogues. European Journal of Medicinal Chemistry. 2015;**89**:13-20. DOI: 10.1016/j.ejmech.2014.10.031

[55] Nandha B, Nargund LVG, Nargund SL, Nandha B. Design and synthesis of some new imidazole and 1,2,4-triazole substituted fluorobenzimidazoles for antitubercular and antifungal activity. Der Pharma Chemcia. 2013; **5**(6):317-327

[56] Birajdara SS, Girish D, Ashish PK, Kamble VM. Synthesis and biological evaluation of amino alcohol derivatives of 2-methylbenzimidazole as antitubercular and antibacterial agents. Journal of Chmial and Pharmacutical Research. 2013;**5**(11):583-589

[57] Anand N, Ramakrishna KKG, Gupt MP, Chaturvedi V, Singh S, Srivastava KK, et al. Identification of 1-[4-(benzyloxyphenyl)-but-3-enyl]-*1H*-azoles as new class of anti tubercular and antimicrobial agents. ACS Medicinal Chemistry Letters. 2013;**4**(10):958-963. DOI: 10.1021/ml40
02248

[58] Awasthi D, Kumar K, Knudson SE, Slayden RA, Ojima I. SAR studies on trisubstituted benzimidazoles as inhibitors of Mtb FtsZ for the development of novel antitubercular agents. Journal of Medicinal Chemistry. 2013;**56**(23):9756-9770. DOI: 10.1021/jm401468w

[59] Keng Yoon Y, Ashraf Ali M, Choon TS, Ismail R, Chee WA, Suresh

Kumar R, et al. Antituberculosis: Synthesis and antimycobacterial activity of novel benzimidazole derivatives. BioMed Research International. 2013; **2013**:1-6. DOI: 10.1155/2013/926309

[60] Ranjith PK, Rajeesh P, Haridas KR, Susanta NK, Guru Row TN, Rishikesan R, et al. Design and synthesis of positional isomers of 5and 6-bromo-1-[(phenyl)sulfonyl]-2-[(4-nitrophenoxy) methyl]-*1H*-benzimidazoles as possible antimicrobial and antitubercular agents. Bioorganic and Medicinal Chemistry Letters. 2013;**23**(18):5228-5234. DOI: 10.1016/j.bmcl.2013.06.072

[61] Patel RV, Patel PK, Kumari P, Rajani DP, Chikhalia KH. Synthesis of benzimidazolyl-1,3,4-oxadiazol-2ylthio-*N*-phenyl (benzothiazolyl) acetamides as antibacterial, antifungal and antituberculosis agents. European Journal of Medicinal Chemistry. 2012;**53**: 41-51. DOI: 10.1016/j.ejmech. 2012.03.033

[62] Sangani CB, Jardosh HH, Patel MP, Patel RG. Microwave-assisted synthesis of pyrido [1,2-a]benzimidazole derivatives of β-aryloxyquinoline and their antimicrobial and anti tuberculosis activities. Medicinal Chemistry Research. 2012;**22**(6):3035-3047. DOI: 10.1007/s00044-012-0322-5

[63] Gobis K, Foks H, Bojanowski K, Augustynowicz-Kopec E, Napiórkowska A. Synthesis of novel 3-cyclohexylpropanoic acid-derived nitrogen heterocyclic compounds and their evaluation for tuberculostatic activity. Bioorganic & Medicinal Chemistry. 2012;**20**(1):137-144. DOI: 10.1016/j.bmc.2011.11.020

[64] Saleshier FM, Suresh S, Anitha N, Karim J, Divakar MC. Design, docking and synthesis of some 6-benzimidazoyl pyrans and screening of their anti tubercular activity. European Journal of Experimental Biology. 2011;**1**(2):150-159

[65] Camacho J, Barazarte A, Gamboa N, Rodrigues J, Rojas R, Vaisberg A, et al. Synthesis and biological evaluation of benzimidazole-5-carbohydrazide derivatives as antimalarial, cytotoxic and antitubercular agents. Bioorganic and Medicinal Chemistry. 2011;**19**(6): 2023-2029. DOI: 10.1016/j.bmc. 2011.01.050

[66] Kumar K, Awasthi D, Lee SY, Zanardi I, Ruzsicska B, Knudson S, et al. Novel trisubstituted benzimidazoles, targetingmtbftsz, as a new class of antitubercular agents. Journal of Medicinal Chemistry. 2011;**54**(1): 374-381. DOI: 10.1021/jm1012006

[67] Pieroni M, Tipparaju SK, Lun S, Song Y, Sturm AW, Bishai WR, et al. Pyrido[1,2-a]benzimidazole-based agents active against tuberculosis (TB), multidrug-resistant (MDR) TB and extensively drug-resistant (XDR) TB. ChemMedChem. 2011;**6**(2):334-342. DOI: 10.1002/cmdc.201000490

Chapter 4

Exploring the Versatility of Benzimidazole Scaffolds as Medicinal Agents: A Brief Update

Gopakumar Kavya and Akhil Sivan

Abstract

Benzimidazole, one of the finest classes of heterocyclic aromatic compounds have the characteristic structure of benzene fused with a five-membered imidazole ring. Despite being made their first appearance in the late 1870s, they are considered as a 'privileged molecule'. The applications of this wonder molecule range from medicinal chemistry to material science. Benzimidazole being a potent inhibitor for various enzymes has got therapeutic effects like anticancer, antimicrobial, anthelmintic, antioxidant, anticonvulsant, antifungal, anti-inflammatory, antiviral, antihistaminic, antipsychotic, etc. It has also made its existence in various branches of medical science *viz* ophthalmology, neurology, cardiology and more. The applications of benzimidazole are not only limited to the biological field but also expanded to the field of material chemistry as well. This chapter summarizes the pharmacological properties of benzimidazole, illustrated on numerous derivatives since 2016.

Keywords: anti-cancer agent, Benzimidazole, biological activity, N-heterocycle, medicinal chemistry, pharmacophore

1. Introduction

The benzimidazole nucleus is fairly unique among heterocyclic ring systems because of its outstanding structural similarity with various naturally occurring nucleotides [1]. In 1872, Hoebrecker synthesized the first benzimidazole molecule by the reduction of 2-nitro-4-methylacetanilide [2]. The biological significance is because its structure is similar to purines, and the importance of the applications depends on their abundance in most of the biologically active molecules. The discovery of the structure of vitamin B_{12} with 5, 6-dimethylbenzimidazole moiety in it, also elicited the search for benzimidazole - similar motifs for various pharmacological applications [3–5]. Following this, various research groups have outlined the synthesis and applications of benzimidazole [6, 7]. Benzene when fused with imidazole results in the formation of benzimidazole (**1**), which can readily undergo tautomerization as shown in **Figure 1**.

The greater reactivity of the 2nd position towards various electrophiles and nucleophiles is the outcome of tautomerization. Many drugs contain the benzimidazole

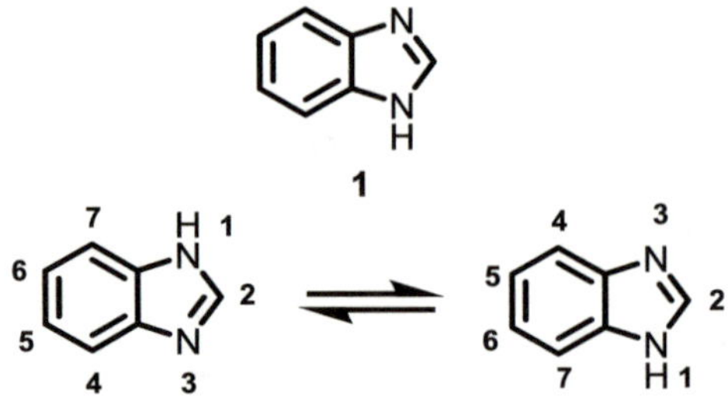

Figure 1.
Benzimidazole (compound 1) with its tautomeric forms.

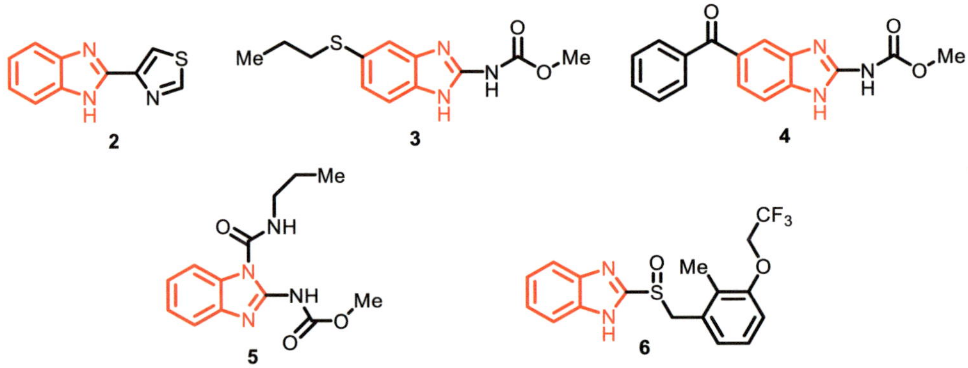

Figure 2.
Drugs (compound 2, 3, 4, 5 and 6) based on benzimidazole core.

nucleus as a core unit and have a widespread application in the pharmaceutical field [8–14]. The presence of benzimidazole pharmacophore in the various branches of medical science is inexplicable. The therapeutic uses of benzimidazole include anti-cancer [15–20], antimicrobial [21–24], antiparasitic [25, 26], anti-inflammatory and analgesics [27–29], antiviral [30–32], and antiulcerative [33] activities and in fields like ophthalmology, neurology, endocrinology, etc. The first example of a benzimidazole that was clinically available was thiabendazole (**2**), which can be used as a fungicide and for antiparasitic purposes [34]. The 2, 6-disubstituted albendazole (**3**) and mebendazole (**4**) were used as anthelmintic or antiparasitic agents. The 1, 2-disubstituted benomyl (**5**) was shown to have antifungal and anticancer activities whereas 2-substituted lansoprazole (**6**) acted as a therapeutic agent for the reduced production of stomach acid and cardiac failures (**Figure 2**).

In this chapter, a plethora of benzimidazole analogs with different pharmacological properties such as anticancer, antibacterial, antifungal, antiviral, anticoagulant, anti-inflammatory, antiparasitic, anthelmintic activity, etc. has been discussed.

2. Benzimidazole and its pharmacological significance

Benzimidazoles were initially used as a plant fungicide and veterinary anthelminthic. After the discovery and use of thiabendazole (**2**), these benzimidazole motifs were used in human beings as well. Since then, a wide variety of molecules having the core structure as (**1**) were synthesized and found their application in the medical world as well as in the material domain.

Various substituted derivatives of (**1**) were showcased diversified therapeutic properties such as antiparasitic, anticancer, anthelmintic, antiproliferative, antioxidants, antimicrobials, anti-inflammatory, antivirals, anticoagulants, antihypertensive, anticonvulsant, antidiabetic, lipid level modulators, anti-HIV, immunomodulators, hormone modulators, proton pump inhibitors and antidepressants. They have also used a building block for various other therapeutic agents. Let us have a peep at some of the innumerable reports of pharmacological activities of benzimidazole.

2.1 Anticancer activity

Cytotoxicity of benzimidazole derivatives is well known and, recently Noha et al. reported compounds of benzimidazole which are *N*-(benzimidazothiazolone) acetamides (7) [35]. In vitro analyses provided the cytotoxic activity of (7) over HCT-116 colon cancer cells. The further detailed study delivered the topoisomerase I-β (Topo I-β) and inhibiting activities against tubulin (**Figure 3**).

The role of benzimidazole analogs as potential metal-based DNA-sensor is unanimous. Fluorogenic differential/sequential Schiff base chemosensors which solely consists of benzimidazole derivatives (**8**), for detecting Cu^{2+}, CN^{-}, $P_2O_7^{4-}$, and Zn^{2+} ions in human cervical (HeLa) and breast cancer (MDA-MB-231 and MCF-7) cell lines were designed by Anbu et al. [36] (**Figure 3**).

Drug repurposing of benzimidazole compounds is generally considered for the reason that, it has antitumor activities. Florio and coworkers screened anthelmintics which are derivatives of benzimidazole [37]. Certain drugs like albendazole (**3**), flubendazole (**9**), oxibendazole (**10**) etc. are subjected to the evaluation of their pharmacokinetics and physicochemical properties (**Figure 3**). For the potential repurposing of the drugs in cancer therapy, a silico target prediction was used to access the pharmacology of these benzimidazole compounds.

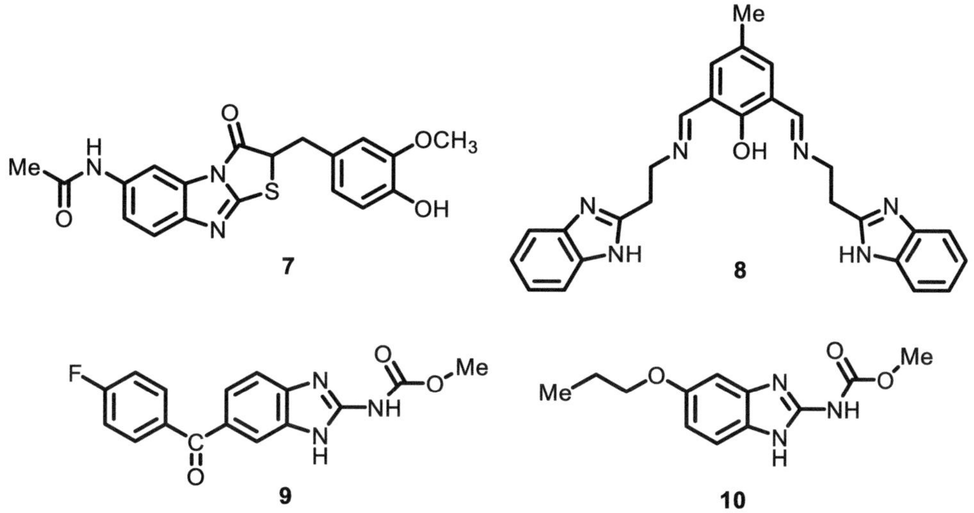

Figure 3.
Benzimidazole derivatives (compound 7, 8, 9 and 10) acts as cytotoxic agents, chemosensors and repurposed drugs.

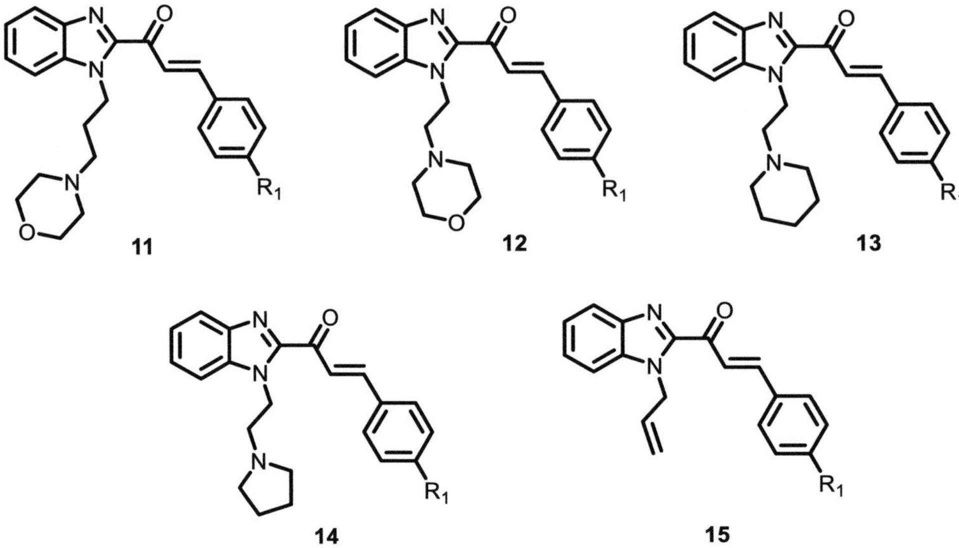

Figure 4.
N-*substituted benzimidazole derivatives (compound* **11, 12, 13, 14** *and* **15***) with antiproliferative activity.*

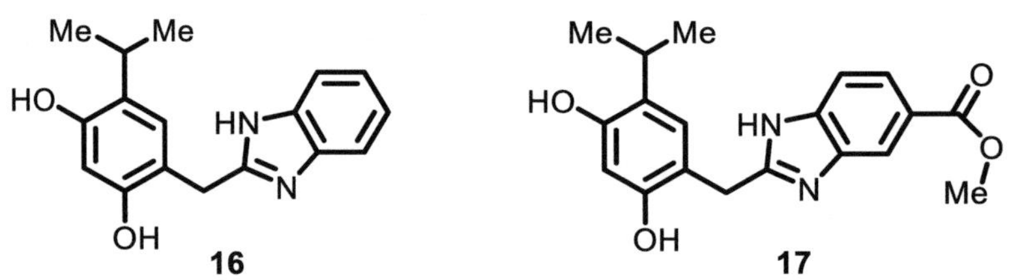

Figure 5.
Benzimidazole derivatives (compound **16** *and* **17***) for the inhibition of HSP90.*

Synthesis of *N*-substituted benzimidazole analogues (**11–15**) with an alkyl chain and a nitrogen-containing 5- or 6-membered ring increased the anticancer effects on human ovarian carcinoma (OVCAR-3) and human breast adenocarcinoma (MCF-7) cell lines, were reported by Hsieh et al. [38]. (2*E*)-1-(1-(3-morpholinopropyl)-1*H*-benzimidazol-2-yl)-3-phenyl-2-propen-1-one) (**11**) acts as the most potent antiproliferative drug and has got more advantages than the standard drug, cisplatin (**Figure 4**).

The stabilization of proteins in the cell is being coordinated by heat shock proteins (HSPs). HSP90 plays a major role in it. This can be reflected in cancer therapy. Neverdauskas et al. synthesized benzimidazole derivatives with resorcinol (**16**) and (**17**), as potential inhibitors for HSP90 (**Figure 5**) [39].

Benzimidazole is considered a privileged molecule in the medicinal world. Hernández-Romero et al. in 2021 synthesized first-row transition metal compounds which contain benzimidazole moieties (**18–21**) as ligands in them (**Figure 6**) [40]. The advancement of metallodrugs for the treatment of cancer has been rapidly evolving. The use of benzimidazole as mono-, di-, tri-, and tetradentate ligands with metals like Cu, Co, Zn, Ni, Mn, V, and Fe led to the formation of effective drugs for cancer therapy by increasing the cytotoxic and antiproliferative activity.

Bistrović et al. synthesized monocationic benzimidazoles (**22**) and (**23**), starting from *o*-phenylenediamines and benzaldehydes having 1,4-disubstituted-1,2,3-triazole

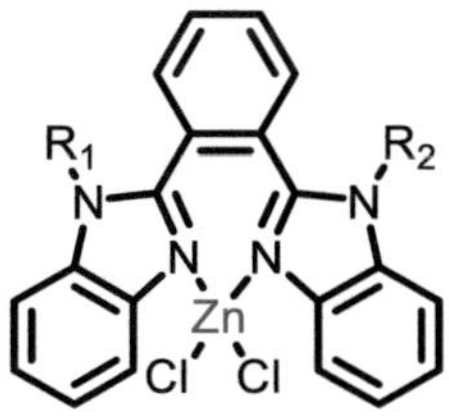

18

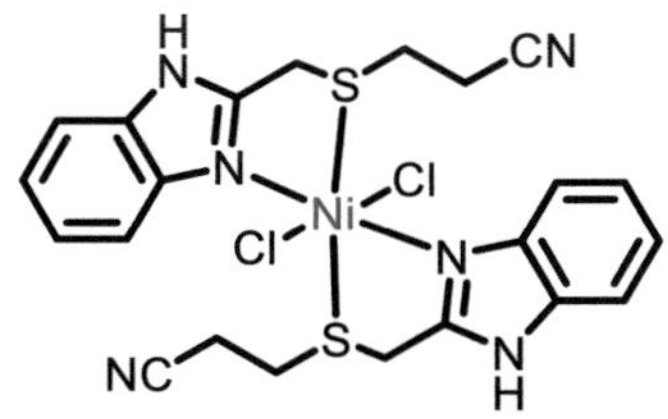

19

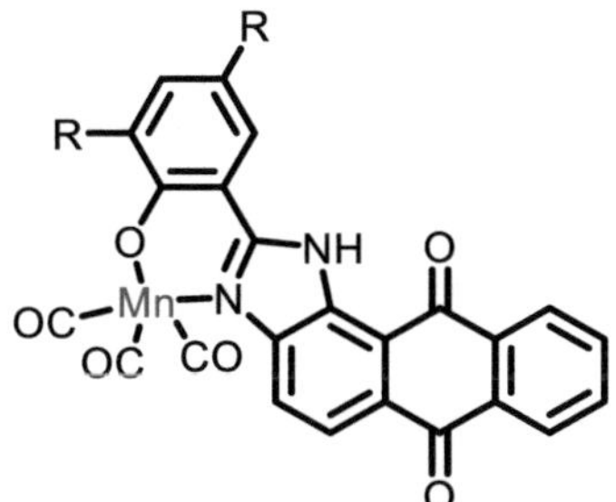

20

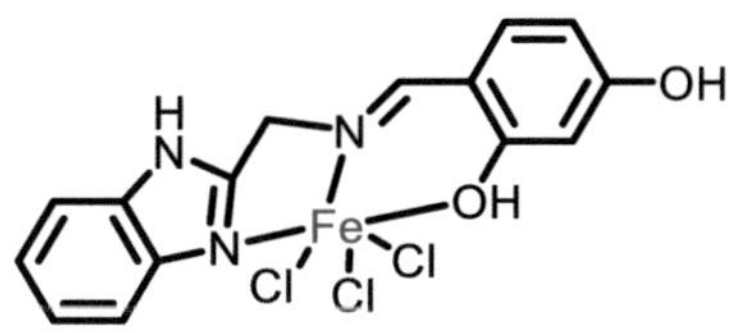

21

Figure 6.
*Benzimidazole derivatives (compound **18**, **19**, **20** and **21**) as metallodrugs.*

motifs and studied its antiproliferative activities [41]. These compounds showed potent and selective activities that are cytostatic against non-small cell lung cancer (A549) in the low nM range and could be because of apoptosis and primary necrosis (**Figure 7**). Because of the presence of different amidino groups and aromatic substituents, these compounds showed a difference in their cytostatic activities in

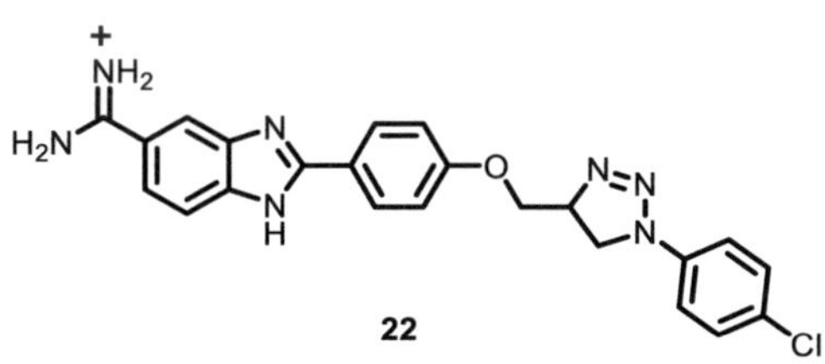

22

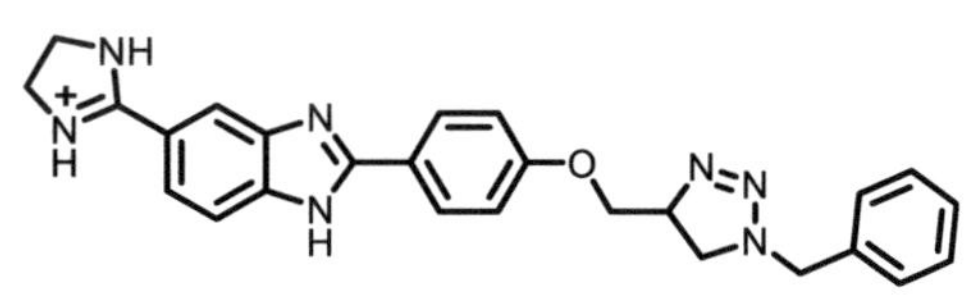

23

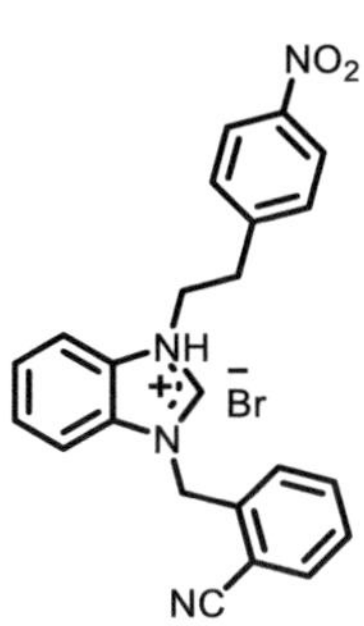

24

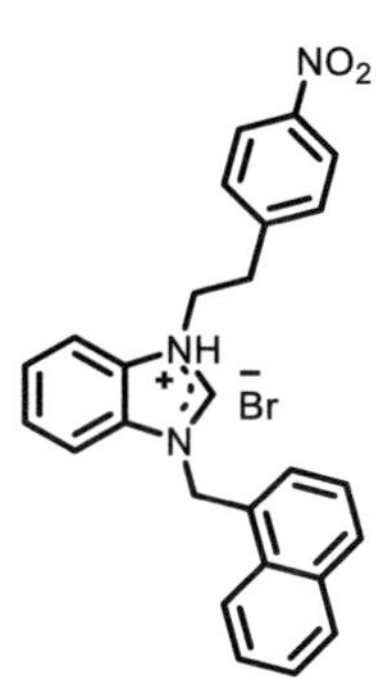

25

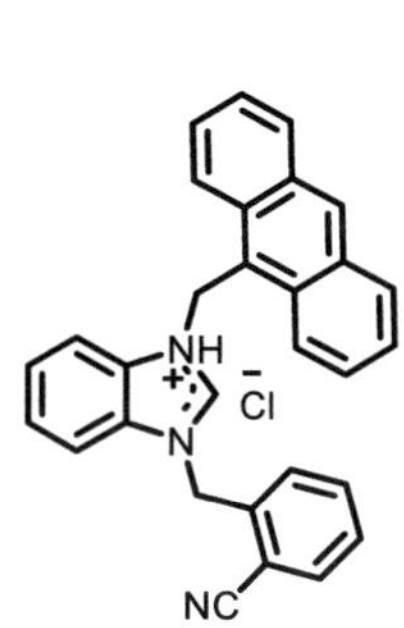

26

Figure 7.
*Benzimidazole derivatives (compound **22**, **23**, **24**, **25** and **26**) with potential inhibition for lung, colon, and breast cancer.*

Western blot analysis. The enzyme p38 MAPK got inhibited by both the compounds as shown by *in silico* structural analysis.

The 1, 3-disubstituted benzimidazoles (**24–26**) were synthesized starting from *o*-diphenylamine and their interactions with cancer cells proteins were examined using molecular docking studies (**Figure 7**) [42]. The biochemical assay of the synthesized compounds were compared by doing theoretical calculations whereas its biological activity was tested against proteins such as colon cancer antigen (ID 2HQ6) and breast cancer (ID 2AR9) by using molecular docking studies. These benzimidazolyl halides were found to be better against the protein molecules studied and of which (**26**) was found to be more potent in action among the given three.

Synthesis of imidazo[1,2-*a*]pyrazine appended benzimidazoles (**27**) and (**28**) was done, starting from 1,3-dibromobenzene and evaluated its anticancer activities on the inhibition of growth of NCI-60 human cancer cell lines [43]. The antiproliferative activity of these molecules is attributed to causing damage to the DNA of such cells. The planar geometry of these compounds also enhanced the intercalated binding with cancer cells DNA. The cytotoxicity evaluation of the compounds was also done against the human normal cell line (Hek293) and found to be very low with higher LC_{50} values (**Figure 8**).

Srour et al. in 2020 reported the formation of a novel class of 2-thiazol linked benzimidazoles (**29**) and studied its inhibiting action against epidermal growth factor receptor (EGFR) (**Figure 8**). The *in vitro* studies of the synthesized compounds using erlotinib as a standard drug revealed its suppression activity against EGFR PK inhibitors, which targets human breast cancer (MCF-7) cells. They have also exhibited a

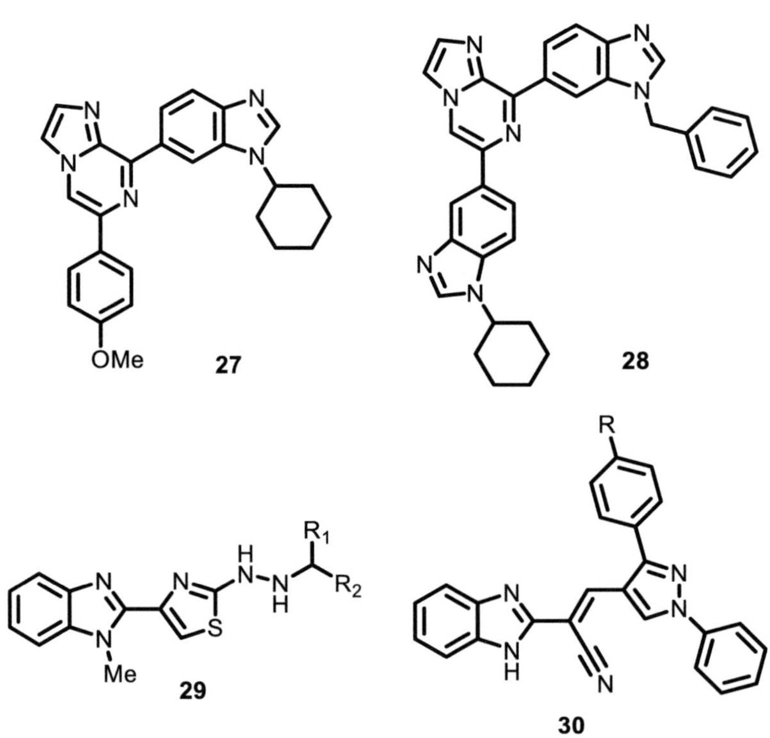

Figure 8.
Benzimidazole derivatives (compound 27, 28, 29 and 30) with NCI-60 cell line inhibition.

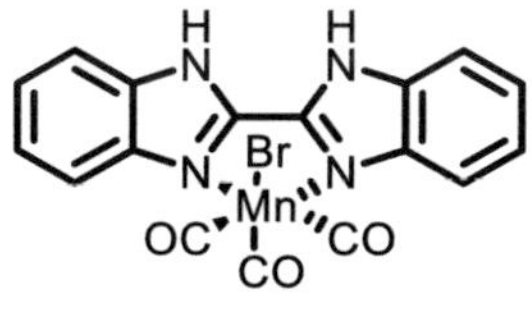

31

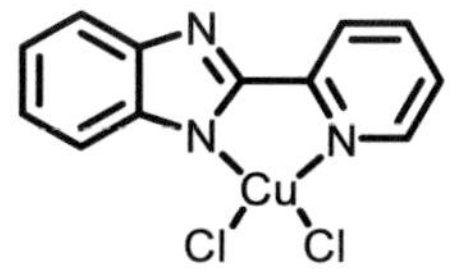

32

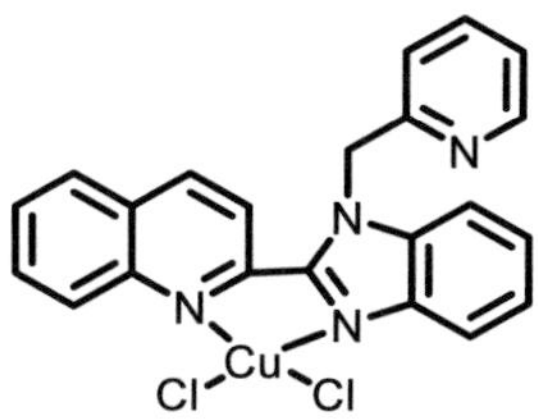

33

Figure 9.
Derivatives of benzimidazole (compound 31, 32 and 33) with anticancer activity against liver, lung, and gastric cancer cells.

very low suppression percentage among normal cells indicating its diminished side effects when used as an antiproliferative drug [44].

Benzimidazole-tethered pyrazoles (**30**) have been synthesized in multi-steps by the condensation of phenylhydrazine with acetylphenones followed by cyclization, Vilsmeier-Haack formylation and Knoevenagel reactions (**Figure 8**) [45]. A study of anti-inflammatory and antioxidant activities of the benzimidazoles showed a marked improvement when compared with diclofenac sodium and ascorbic acid as standards respectively. The anticancer activity was shown against human pancreatic cancer cell line AsPCl (progenitor) and SW1990 (squamous) which was also visible in the better binding with B-cell lymphoma in docking studies.

Mn(I) and benzimidazole co-ligands (**31**) with potential photo-activated carbon monoxide releasing molecules (CORMs) were synthesized and their biological activities were studied (**Figure 9**) [46]. The CO releasing properties, as well as luminescence intensities of these complexes, differed with the extend of conjugation and with the degree of unsaturation present in the benzimidazole co-ligands. The bioimaging capabilities of these complexes were proved by the absorption of it by liver cancer cells (SK-Hep1) and human liver cells (HL-7702) under cellular fluorescence imaging tests. Complex (**31**) showed excellent anticancer activities among all the molecules synthesized.

Prosser et al. synthesized a Cu(II) complex of benzimidazole (**32**) and studied their anticancer properties [47]. This derivative that was revised at the non-coordinated nitrogen of the benzimidazole molecules, exhibited excellent cytotoxicity against A549 adenocarcinomic alveolar basal epithelial cells (**Figure 9**).

Research works concentrating on the effective therapeutic agent possessing antiproliferative activity for human gastric cancer paved the way to the discovery of yet another benzimidazole derivative (**33**) with quinoline copper-based complex (**Figure 9**) [48]. The complex ensures G2/M phase arrest, apoptosis, mitochondrial dysfunction etc. and thus provides effective cytotoxicity.

Aromatase inhibitors (AIs) are compounds that control estrogen-related diseases and hence breast cancer, as its concentration was found to be higher in such cases. Çevik et al. in 2020 synthesized some novel benzimidazole- triazolothiadiazine

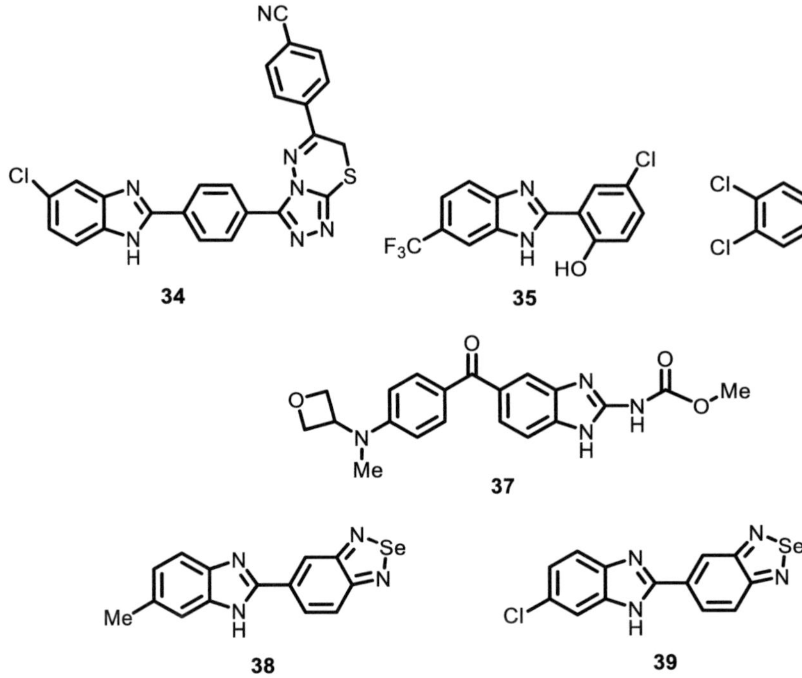

Figure 10.
Benzimidazole derivatives (compound 34, 35, 36, 37, 38 and 39) provide inhibition against breast, lung, and prostate cancer cells.

libraries and examined its aromatase inhibition activities [49]. Initial screening of these compounds towards anticancer properties against breast cancer cell line (MCF-7) in humans, resulted in getting good results. Upon further subjecting it to *in vitro* aromatase enzyme inhibition studies, the compound (**34**).

among them was found to be almost equal in activity when compared with a reference drug letrozole (**Figure 10**).

The role of benzimidazole compounds in the treatment of breast cancer is exemplary. Gangrade et al. demonstrated the use of benzimidazole derivatives in the inhibition of Wnt/β-catenin signaling [50]. The upregulation of Wnt/β-catenin signaling in triple-negative breast cancer (TNBC), when compared to normal and other breast cancer subtypes, is inevitable. Benzimidazole compounds like SRI33576 (**35**) and SRI35889 (**36**) have a high cytotoxicity rate in TNBC cell lines. They are found to be active inhibitors of Wnt/β-catenin signaling and have therapeutic properties for treating TNBC (**Figure 10**).

Cheong and co-workers designed and synthesized benzimidazole methylcarbamate analogue (**37**) with enhanced water solubility [51]. The existed drugs that account for the treatment of metastatic cancers are not suitably aiding the circumstances. Poorly soluble benzimidazole methylcarbamate drugs, which are effective anthelmintics are subjected to functionalization with oxetane or an amine group to improve the solubility and then used as an active therapeutic agent for the treatment of metastatic cancers. Cytotoxicity towards prostate, lung, and ovarian cancers is exhibited by the novel oxetanyl substituted compound (**37**) (**Figure 10**).

Liang and co-workers synthesized selenium-containing benzimidazole derivatives through condensation of peptide coupling reagents and irradiation of microwaves [52]. These selenediazole derivatives were recognized as potent anticancer agents against MDA-MB-231 and MCF-7 breast cancer cell lines. Compounds (**38**) and (**39**) showed greater cytotoxic activity towards triple-negative breast cancer cell line MDA-MB-231 (**Figure 10**).

Husain et al. prepared various derivatives of furanone appended benzimidazoles, which effectively contribute to cancer therapy [53]. Compound (**40**) was found active against DU145 and MCF7 whereas compound (**41**) has got excellent activity against MCF7, A549, and DU145 cell lines (**Figure 11**). They are potential cytotoxic agents than the standard drug doxorubicin.

Compounds (**42**) and (**43**) are *bis*-benzimidazole analogs that have been synthesized to account for cancer therapy under microwave irradiation [54]. The anticancer activity was studied with the help of.

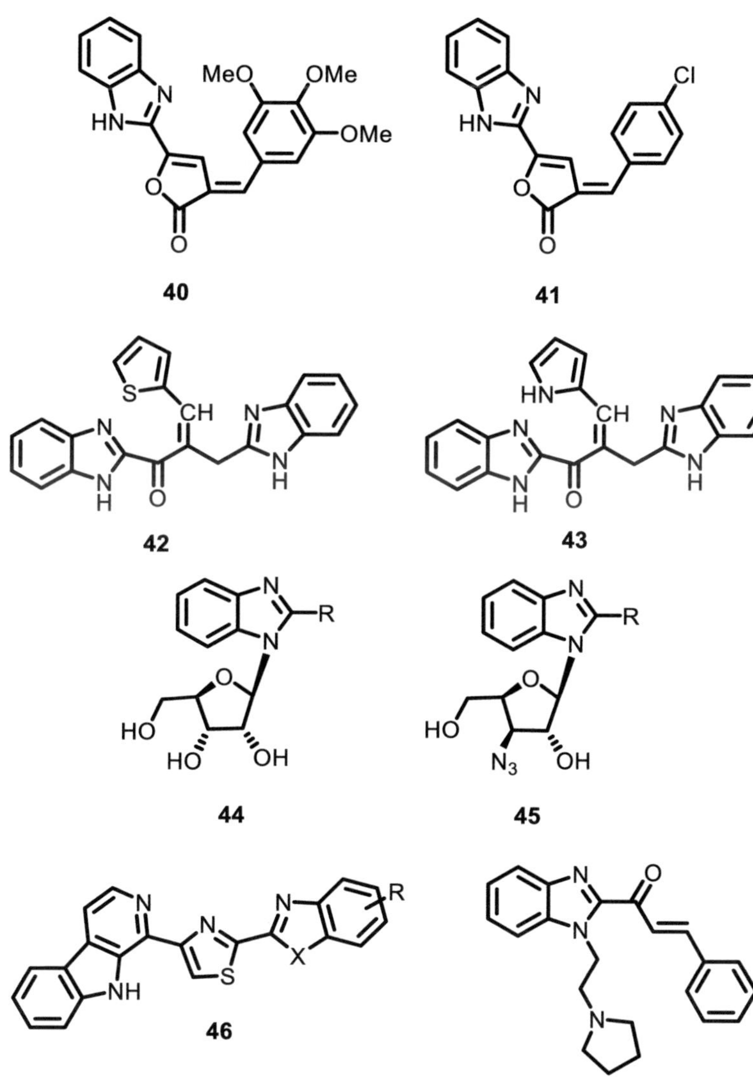

Figure 11.
Benzimidazole analogs (compound 40, 41, 42, 43, 44, 45, 46 and 47) with anticancer activities.

Molinspiration software and they possess high bioactivity scores. It was also found that they obey Lipinski's rule and could be emerged as a lead anticancer drug (**Figure 11**).

Shinde and co-workers used D-glucose as the precursor for the synthesis of ribofuranosyl nucleosides (**44**) and (**45**) (**Figure 11**). Evaluation of their anticancer activity was done using the MDA-MB-231 cell line [55].

Sireesha et al. designed and synthesized benzimidazole/benzoxazole-linked β-carbolines (**46**) by the condensation of two various anti-cancer fragments (**Figure 11**) [56]. With the assistance of MTT assay, these compounds were subjected for the anti-cancer screening against Colo-205 (colon), MCF-7 (breast), A2780 (ovarian), and A549 (lung) and found that these exhibits maximum anti-cancer activity with the β-carbolines hybrid.

Benzimidazole derivatives (**47**) with a pyrrolidine side chain can be effectively used to treat sorafenib resistance (SR) in hepatocellular carcinoma, was reported in 2019 [57]. Mode of action is through the inhibition of proliferation of SR cell lines by interrupting the phosphorylation of AKT, p70S6, and the downstream molecule RPS6 (**Figure 11**).

Synthesis of organoruthenium(II) complexes of benzimidazoles (**48**) and (**49**) was reported by Welsh and coworkers (**Figure 12**) [58]. Their anti-cancer activity was screened against triple-negative MDA-MB-231 and MCF-7 breast cancer cell lines, respectively. Among the synthesized compounds, (**48**) showed more potency and (**49**) showed comparable potency with the cisplatin, against the MCF-7 cell line.

2.2 Antibacterial and antifungal activity

Heterocyclic appended benzimidazoles were synthesized and their antibacterial and antifungal activities were tested [59]. The mechanism of action of these molecules was also examined by using docking studies with bacterial proteins such as DNA gyrase subunit B (DNAG) and penicillin-binding protein 1a (PBP1a). The compounds with thiazole and thiadiazole moieties (**50**) and (**51**) respectively, showed marked inhibitory activity against *Escherichia coli*, *Bacillus pumilus*, and *Staphylococcus aureus* bacteria (**Figure 13**).

Ajani et al. synthesized various *o*-substituted and 1, 2-disubstituted benzimidazoles and examined their antibacterial properties [60].

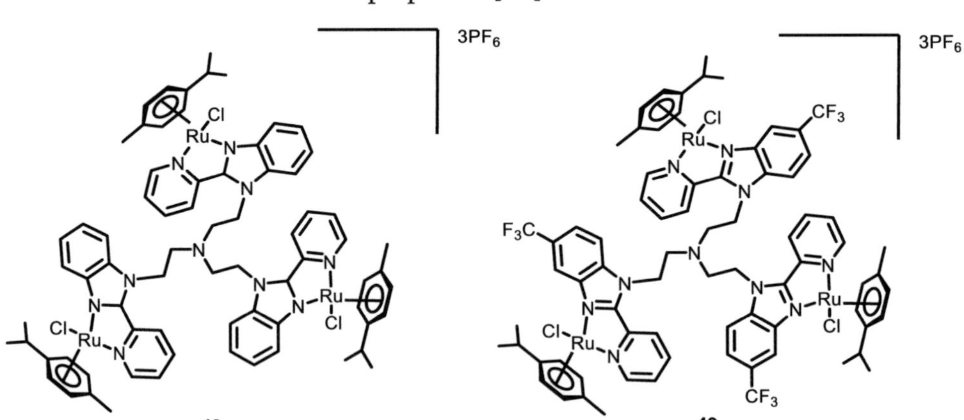

48

49

Figure 12.
*Organoruthenium benzimidazole derivatives (compound **48** and **49**) with inhibition against breast cancer cell lines.*

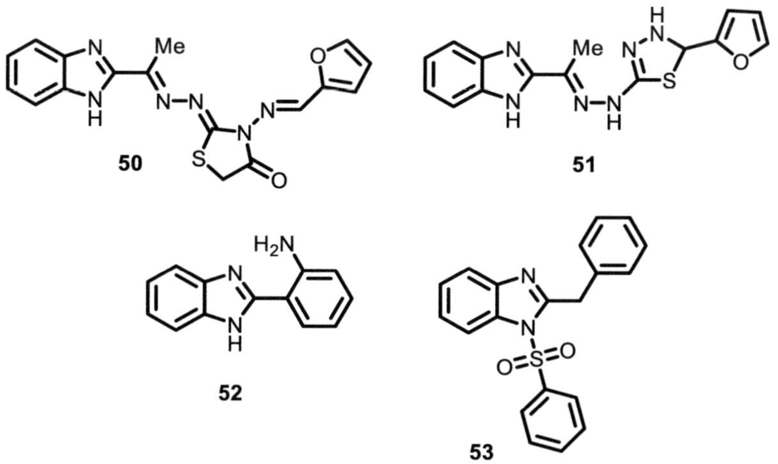

Figure 13.
Benzimidazole derivatives (compound 50, 51, 52 and 53) with inhibitory activity against common bacteria.

Benzene-1,2- diamine undergoes condensation reactions with anthranilic acid, 3, 5-dinitrophenylbenzoic acid, and phenylacetic acid, catalyzed by NH_4Cl yielded the precursor molecules, which on reaction with electrophile-releasing agents produced the corresponding *o*-substituted and 1,2-disubstituted benzimidazoles (**52**) and (**53**), respectively (**Figure 13**). *In vitro* studies of these compounds showed a better activity with a low minimum inhibitory concentration (MIC) value.

1-aryl-substituted 1, 2, 3-triazole appended amidinobenzimidazoles linked *via* phenoxymethylene units (**54**) and (**55**) were synthesized and their anti-bacterial as well as anti-trypanosomal activities and DNA/RNA binding affinities, were studied [61]. Compound (**54**) showed a remarked inhibition against gram-positive bacteria whereas compound (**55**) showed inhibition against gram-negative bacteria. These compounds also showed binding affinities towards ctDNA. Compound (**56**) with *N*-isopropylamidine and *p*-methoxyphenyl-1,2,3-triazole units exhibited enhanced anti-trypanosomal activities against *T. brucei* and reduced toxicity towards mammalian cells (**Figure 14**).

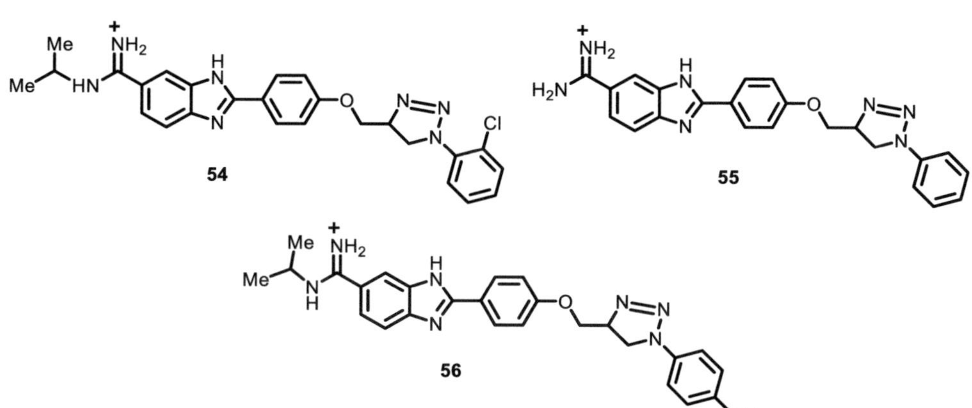

Figure 14.
Benzimidazole derivatives (compound 54, 55 and 56) with antibacterial and anti-trypanosomal activities.

A microwave-assisted, Ni(II) catalyzed novel preparation of 2,6-disubstituted and 1,2,6-trisubstituted benzimidazoles were achieved by Patel and his group (**Figure 15**) [62]. The *in-vitro* antimicrobial studies of the title compounds (**57**) against gram-positive and gram-negative bacteria and fungal strains showed an improved activity exhibited by them when compared with ampicillin, a standard drug. Certain compounds show potent anti-mycobacterium tuberculosis activity, antimalarial activity, antioxidant activity, etc. All these activities were supported by better molecular docking scores and their pharmacokinetics were also examined by ADME-Tox descriptors.

A comparative antimycobacterial activity study of 2,5-disubstituted and 1,2,5-trisubstituted benzimidazoles was reported in 2020 [63]. The *in vitro* studies against *Mycobacterium tuberculosis* H37Rv strain revealed an increased activity correlated with lipophilicity for disubstituted compounds (**58–60**) than for trisubstituted ones because of the addition of a long hydrocarbon chain at position 1 in the latter (**Figure 15**).

A library of mono and disubstituted benzimidazoles were synthesized by applying different methodologies, i.e., by using the microwave, ultrasound (US), infrared (IR), simultaneous application of US and IR, and by conventional heating [64]. The antimicrobial and antifungal activities of these benzimidazole derivatives were then evaluated. It was found that some compounds such as (**61**) and (**62**), were proved to be a better substitute than the standard drugs trimethoprim sulfamethoxazole and miconazole for antimicrobial and antifungal activities, respectively (**Figure 16**).

Very recently, Khan et al. designed and synthesized pyrimidine-benzimidazole hybrids (**63**) using the revised Biginelli reaction and evaluated its potential inhibition of SARS-CoV-2 main protease and spike glycoprotein [65]. Investigation about the pharmacological properties resulted in biological evidence like antimicrobial and antifungal properties. The derivatives developed possess more affinity in binding and anti-SARS-CoV-2 activity than presently approved drugs (**Figure 16**).

Zha et al. demonstrated benzimidazole derivatives (**64**) and (**65**) as potent antibacterial agents (**Figure 16**) [66]. Properties like enzyme inhibition, DNA binding, and having a synergistic effect with existing antibiotics makes benzimidazole an active warrior against methicillin-resistance *Staphylococcus aureus* (MRSA).

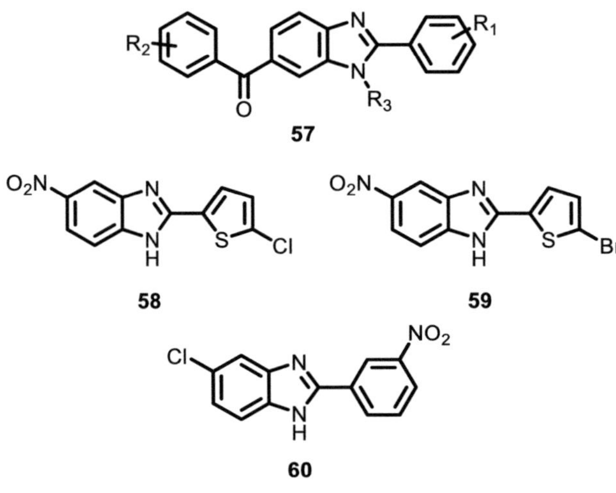

Figure 15.
Benzimidazole derivatives (compound 57, 58, 59 and 60) with antibacterial and antimycobacterial activities.

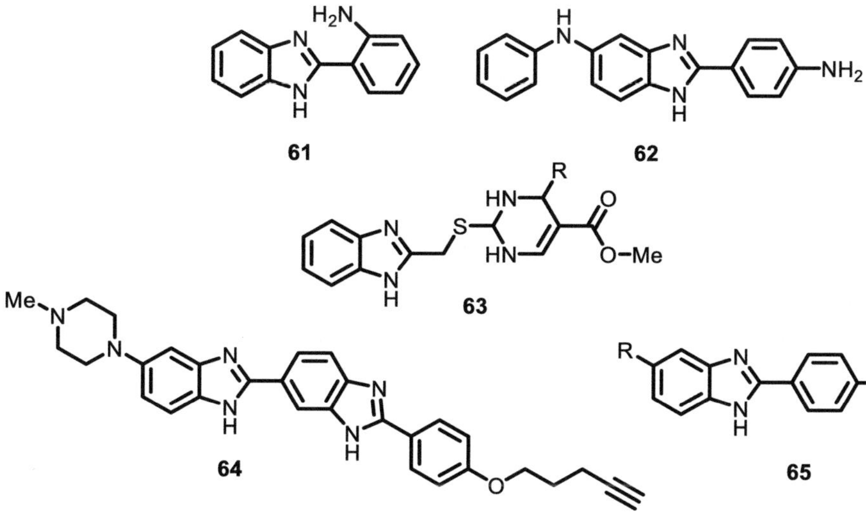

Figure 16.
Potential benzimidazole-derived antibiotics (compound 61, 62, 63, 64 and 65).

Claisen-Schmidt condensation of 2-acetylbenzimidazole and aldehydes followed by a series of steps resulted in the synthesis of benzimidazole derivative (**66**) [67]. They exhibit exceptional anti-microbial and anti-bacterial activities. The grafting of certain functional groups and the presence of pyridine, pyrimidine, indole, etc. improvises the anti-microbial activity (**Figure 17**).

Karaburun et al. described the multi-step synthesis of a series of benzimidazole-1,3,4-oxadiazole derivatives (**67**) which are prominent for their antifungal activities against *Candida* species [68]. The ergosterol inhibition power was proven *via* ergosterol quantification assay and the docking studies were performed on 14-α-sterol (**Figure 17**).

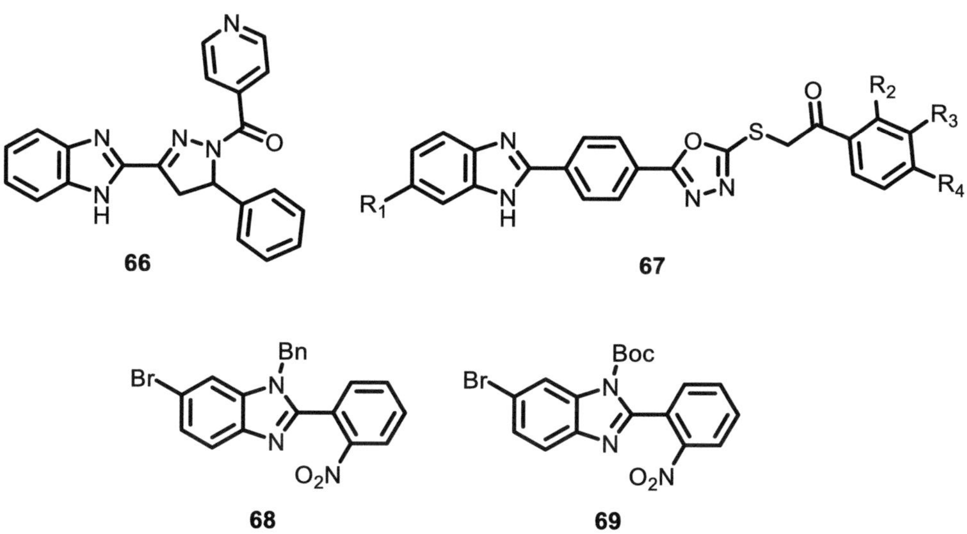

Figure 17.
Benzimidazole derivatives (compound 66, 67, 68 and 69) with antimicrobial, antifungal, and antibacterial activities.

Recently, Aroso and co-workers computationally designed benzimidazole derivatives through palladium-catalyzed reactions [69]. The reaction between 4-bromo-1,2-diaminobenzene and 2-nitrobenzaldehyde, followed by a couple of palladium-catalyzed Suzuki–Miyaura and Buchwald-Hartwig amination cross-coupling reactions resulted in the formation of **(68)** and **(69)** (**Figure 17**). The importance of these benzimidazoles is that it has an inhibitory effect on *E. coli* DNA gyrase B.

Chen et al. designed flavonoid analogs (**70**) which consist of benzimidazole derivatives like 4*H*–chromen-4-one, which provides a remarkable anti-bacterial resistance against members of *Xanthomonas* and *Ralstonia solanacearum* (**Figure 18**) [70]. Molecular docking studies showed the curative and protective activity for the *Tobacco mosaic virus* (TMV). The inhibition rate value is high for these analogs when compared with other anti-viral agents.

Compound (**71**) synthesized by Gençer and co-workers were tested against *Candida* species through microdilution methods [71]. MTT assay and other various microbiological studies provided the antifungal profile with good and effective *in vitro* cytotoxic effects along with inhibition on ergosterol biosynthesis (**Figure 18**).

Synthesis of benzimidazole derivatives like triazinane (**72**) and oxidiazinanes (**73**) through the process of amino methylation with the aid of different aryl-*N*, *N′* unsymmetrical thioureas were designed by Gullapelli and co-workers (**Figure 18**) [72]. The antibacterial activity was evaluated by using suitable gram-positive and negative bacterial strains.

Wang et al. reported the synthesis of a series of benzimidazole moieties (**74–76**) with quinolone analogs which exhibited antibacterial and antifungal properties (**Figure 19**) [73]. The bioactive assay proved that the 2-fluorobenzyl derivative has got remarkable antimicrobial activities against the *P. aeruginosa* and *C. tropicalis*.

A novel, one-pot synthesis of 2-substituted benzimidazoles and Mannich bases (**77–80**) with potent antimicrobial activity was reported by Marinescu et al. [74].

Qualitative and quantitative antimicrobial bioassay of these benzimidazole derivatives showed activity against a broad spectrum of gram-positive and negative bacterial strains both in planktonic and adherent states. The presence of nucleophilic groups like -OH or -CH_3 accounts for the microbicidal activity (**Figure 20**).

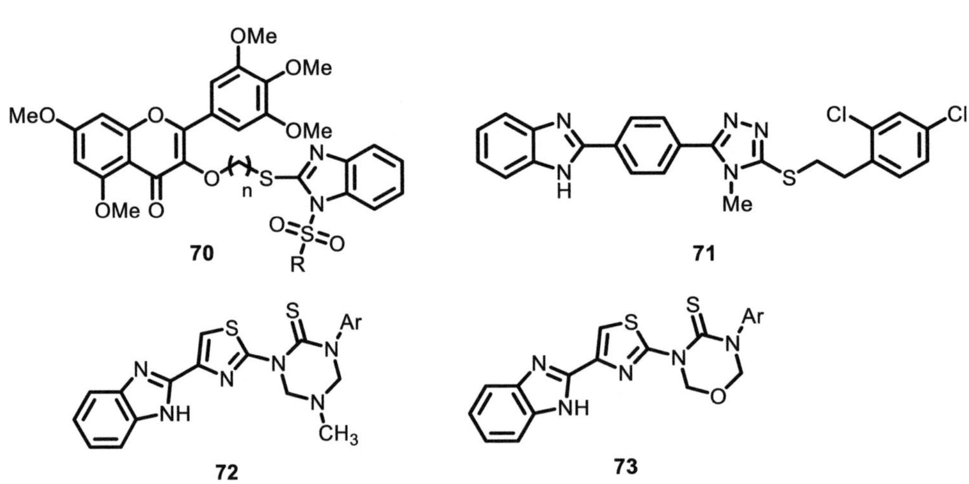

Figure 18.
Derivatives of benzimidazoles (compound 70, 71, 72 and 73) with potential antibacterial, antifungal, and cytotoxic activities.

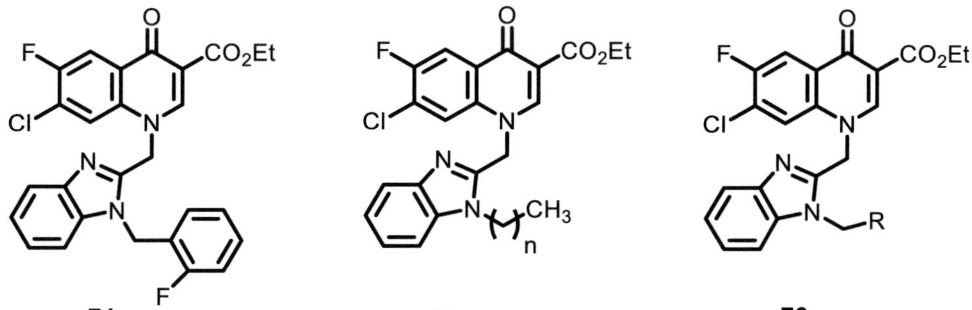

Figure 19.
Quinolone analogs of benzimidazole (compound 74, 75 and 76) with antibacterial and antifungal activities.

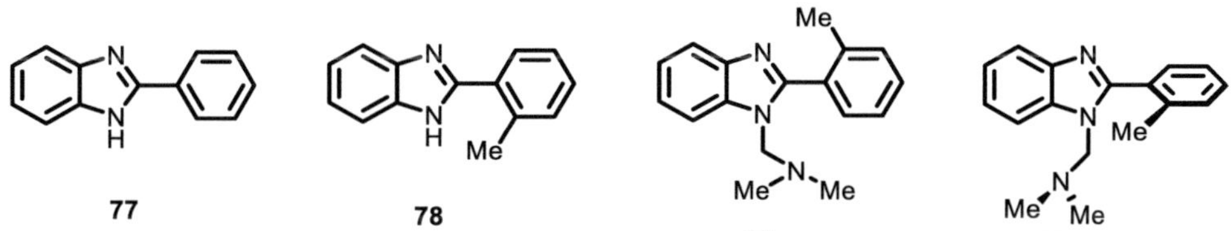

Figure 20.
Potent antimicrobial derivatives of benzimidazole (compound 77, 78, 79 and 80).

Benzimidazoles moieties linked with *N*-acyl substituted indole (**81–86**) were demonstrated by Abraham et al. [75]. The assessment of antimicrobial activity was done against gram-negative and gram-positive bacteria like *Pseudomonas aeruginosa* (MTCC424), *Staphylococcus aureus* (MTCC 2940), *Escherichia coli* (MTCC 443), and *Enterococcus fecalis*. These compounds also account for the hindering of biofilm formation and then the effective growth of *Staphylococcus epidermis* (**Figure 21**). Along with this, an HRBC membrane stabilization test was carried out for the evaluation of the anti-inflammatory activity.

Antoci and co-workers synthesized *bis*-(imidazole/benzimidazole)-pyridine derivatives (**87**) through *N*-alkylation (**Figure 22**) [76]. The anti-TB activity of the compound is good to excellent against both replicating and nonreplicating Mtb. The derivatives are effective against drug-resistant Mtb and some possess a bactericidal approach.

The synthesis of naphthyl-substituted benzimidazole derivatives (**88**) and (**89**) was reported by Ersan et al. in 2020 [77]. The antimicrobial activity was screened and was found that (**89**) showed maximum potency against all gram-positive and gram-negative bacteria. Also, (**88**) actively functions as an antifungal agent. These derivatives also interact with active sites of *E. coli* and can be accounted for inhibition of *E. coli* topoisomerase I (**Figure 22**).

Sirim et al. designed and synthesized benzimidazole-acrylonitrile hybrid derivatives from benzene-1, 2-phenyleneamine and ethyl cyanoacetate followed by reaction with piperazines [78]. All the derived compounds exhibited anti-mycobacterial activity against *M. tuberculosis* H37Rv strain by microplate alamar blue assay (MABA). Compound (**90**) was found to be more effective than standard drugs like isoniazid, ciprofloxacin, rifampicin, etc. (**Figure 22**).

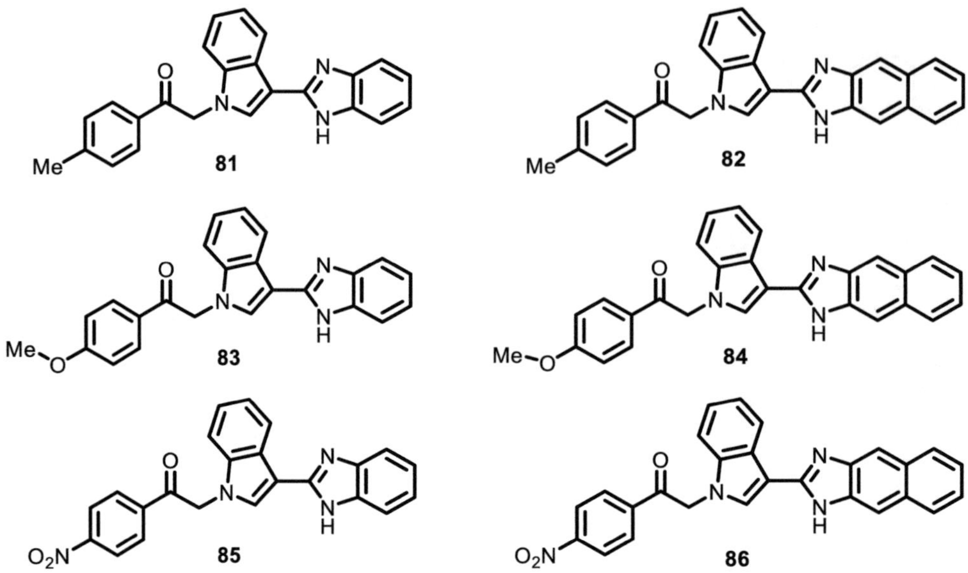

Figure 21.
N-acyl substituted indole-linked benzimidazole derivatives (compound ***81, 82, 83, 84, 85*** *and* ***86****) as antimicrobial agents.*

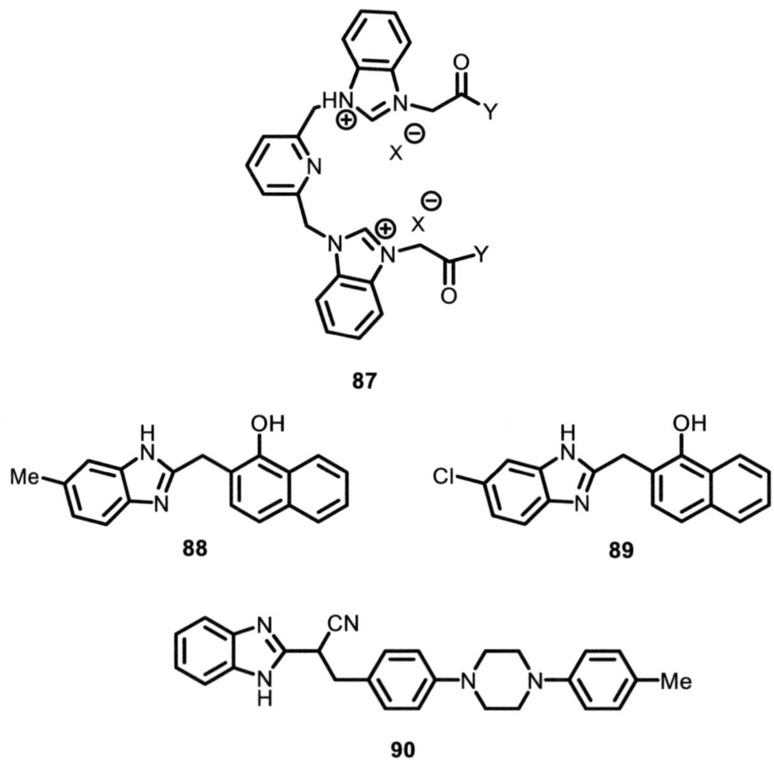

Figure 22.
Benzimidazole derivatives (compound ***87, 88, 89*** *and* ***90****) with antibacterial and anti-TB activities.*

2.3 Antiparasitic activity

Taman et al. evaluated the antischistosomal activity of newly synthesized benzimidazole-related compounds like NBTP-OH (**91**) and NBTP-F (**92**) [79]. The Suzuki-Miyaura coupling reaction of 5-formyl thiophen-2-ylboronic acid and 1-bromo-4-hydroxy benzene or 1-bromo-4-flouro benzene followed by a series of reactions resulted in the formation of compounds (**91**) and (**92**), respectively (**Figure 23**). To date, the treatment of schistosomiasis depended on Praziquantel (PZQ). The use of these two structurally related benzimidazole derivatives can be an alternative for PZQ. The *in vitro* schistosomicidal assay performed on adult worms gave the conclusion that they were considered dead through the destruction of tegument after two minutes of treatment with **91** and **92**.

Synthesis of 1, 3-disubstituted benzimidazol-2-ones (**93**) and (**94**) starting from *o*-phenylenediamine and urea, followed by the evaluation of its anti-trichinellosis efficacy was done [80]. It was found that the synthesized benzimidazole derivatives are more effective than the standard drug albendazole, which is the traditional drug used in the treatment against *Trichinella spiralis*. The estimation of antiparasitic activity was employed through the Campbell method. Selective binding of benzimidazole moiety with the β-tubulin of the parasite results in the destruction of the cell, followed by the death of the parasite. The in vitro activity of all the tested benzimidazole analogs increases with the concentration against the Trichinella spiralis (**Figure 23**).

Molecular docking studies and quantitative structure–activity relationship (QSAR) delivered that benzimidazole derivatives (**95–97**) can be used as cruzain inhibitors for the deadly Chagas disease [81]. The model compounds used displayed a high statistical consistency and a notable capability to predict the inhibiting sites (**Figure 24**). The scenario with Chagas disease is the unavailability of an effective treatment method. Clinical studies related to the Chagas disease and cruzain inhibitors have been on the

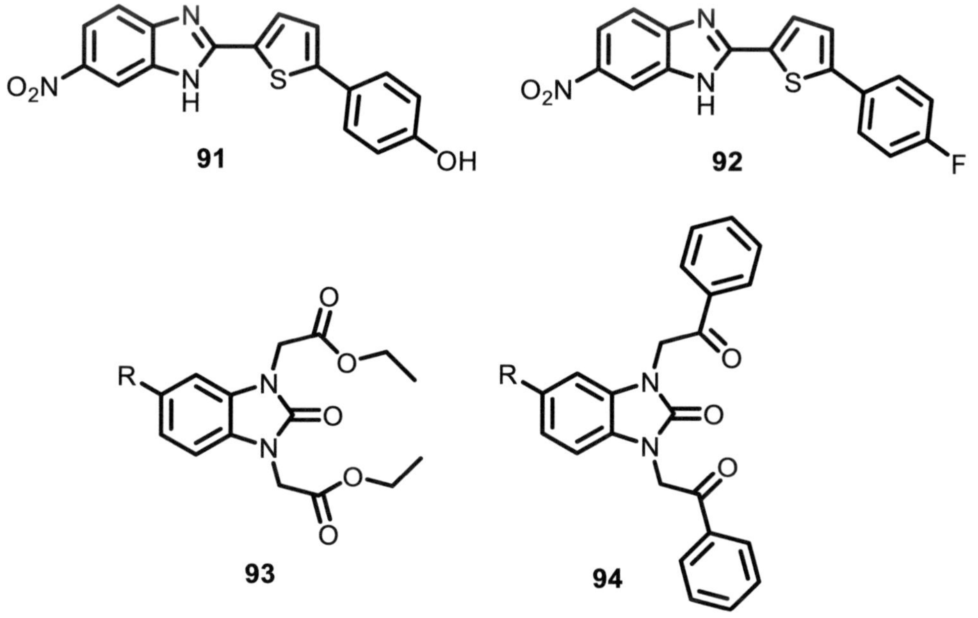

Figure 23.
Benzimidazole derivatives (compound 91, 92, 93 and 94) with antischistosomal and anti-trichinellosis activities.

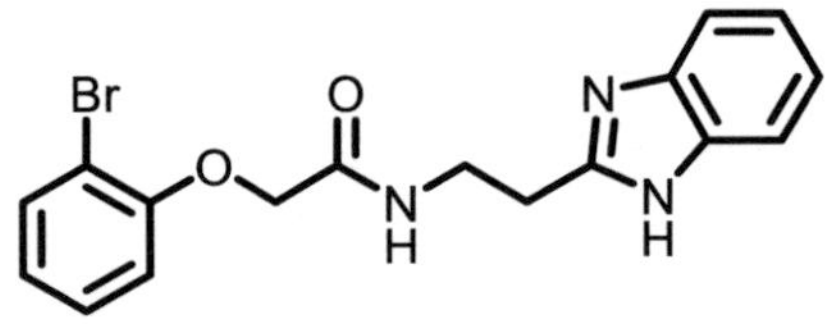

95

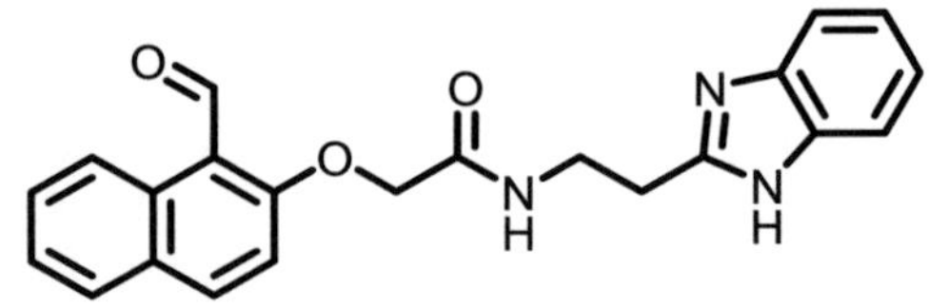

96

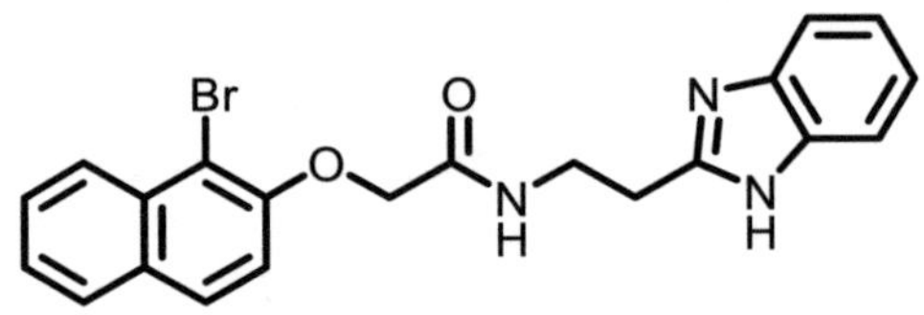

97

Figure 24.
Benzimidazole derivatives (compound 95, 96 and 97) as cruzain inhibitors for Chagas disease.

account of research scientists. In short, benzimidazole derivatives can be used as a lead in the drug discovery of Chagas disease by acting against the recombinant cruzain enzyme.

Tonelli et al. designed and synthesized benzimidazole derivatives from benzene-1, 2-diamine and various acids followed by suitable functionalization and used it as a potent antileishmanial agent [82]. Benzimidazole derivatives were tested against *Leishmania tropica* and *L. infantum* and were found that compounds bearing the derivatives of 1-lupinyl were commonly more active than dialkylaminoalkyl derivatives and compounds (**98**) exhibited the highest potency among the synthesized compounds (**Figure 25**). The observed antileishmanial activity was a result of the interaction of benzimidazole derivatives with acidic components of the cell membrane leading to its destruction.

Exploration of the inhibitory activity of certain benzimidazole compounds like albendazole (**3**), ricobendazole (**99**), oxfendazole (**100**) etc. resulted in

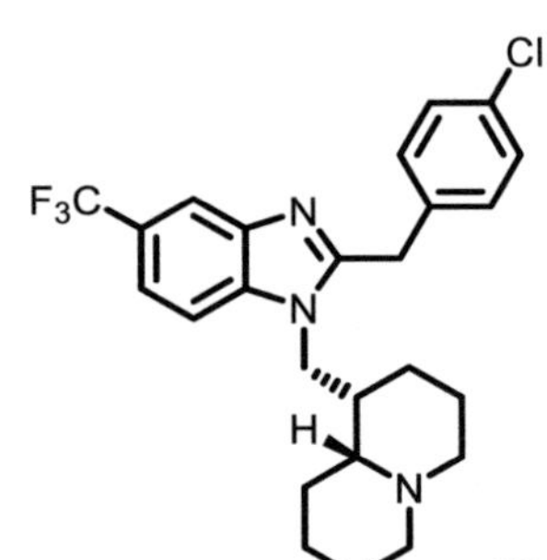

98

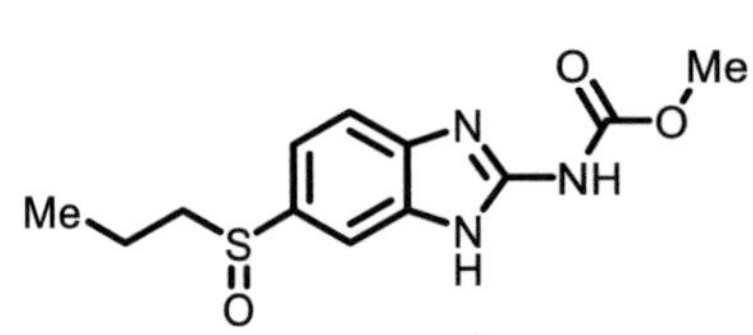

99

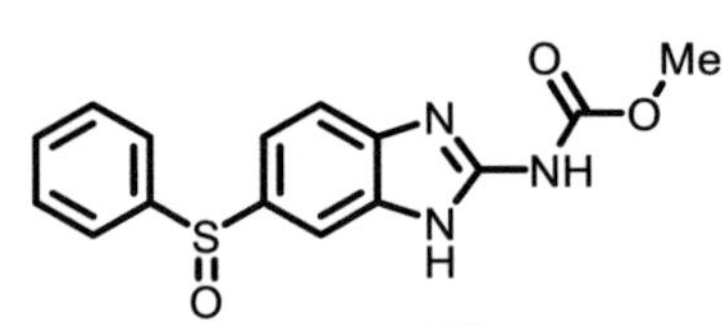

100

Figure 25.
Compounds 98, 99 and 100 with antileishmanial and antiparasitic activities.

acetylcholinesterase (AChE) and butyrylcholinesterase (BChE) inhibition at nanomolar level [83]. This has got immense importance in therapy emerging for handling the resistance among anti-cholinergic factors and in antiparasitic treatment (**Figure 25**). The benzimidazole derivatives are selectively toxic towards helminths which are considered parasites. The inhibitive effect of benzimidazole derivatives on β-tubulin leads to disruption of function in helminths and results in its death.

2.4 Antiviral activity

Compound (**101**) has been synthesized in a multi-step process by Bessieres et al. in 2021 (**Figure 26**) [84]. A study about the inhibition of Ebola virus infection resulted in the design of a more potent and selective drug than the reference drug, Toremifene.

Ibba et al. demonstrated the role of benzimidazole derivatives (**102**) and (**103**), which are active inhibitory agents against *enterovirus* A71 (EV-A71), which is a major cause for foot-mouth disease (HFMD), herpangina, etc. [85]. Penetration and apoptosis assay concluded that the derivatives are capable to inhibit viral endocytosis through reduced viral attachment and penetration to the host cells (**Figure 26**).

Research for the inhibitory action of *chikungunya virus* (CHIKV) infection led to the discovery of benzimidazole-related antiviral agent which targets the nonstructural protein 4 (nsP4), was reported by Wada and co-workers [86]. One of the compounds (**104**), synthesized by them can effectively inhibit CHIKV by using M2295 residue in the nonstructural protein 4 (nsP4) and with the help of CHIKV replicons, it inhibits the RNA-dependent RNA-polymerase (RdRp) function of CHIKV (**Figure 26**).

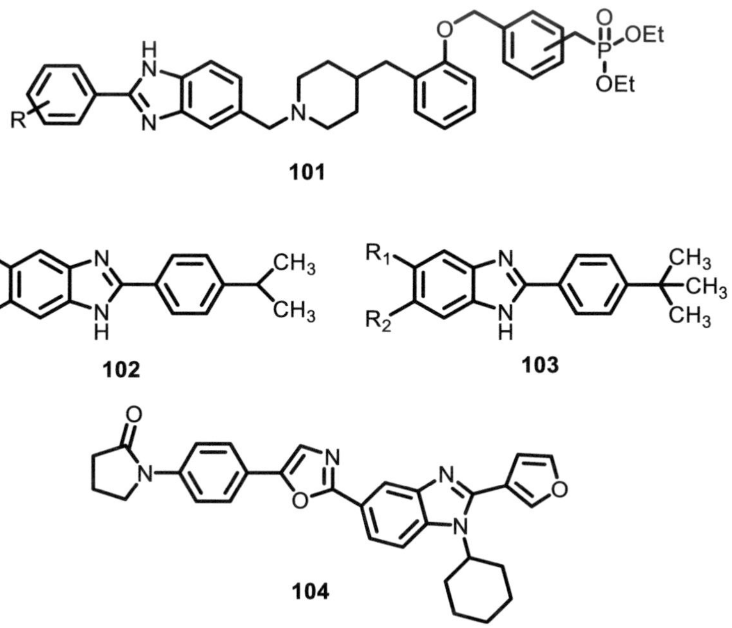

Figure 26.
*Potential antiviral derivatives of benzimidazole (compound **101**, **102**, **103** and **104**).*

2.5 Other properties like antipsychotic, antidiabetic, anticoagulant activities, etc.

In vitro and *in vivo* characteristic studies of benzimidazole acetamide derivatives (**105**) in the ethanol-induced neuro-degeneration model was performed by.

Imran et al. in 2021 [87]. The derivatives lowered the neurodegeneration and inflammation of neurons by down-regulating inflammatory cascades caused by oxidative stress (**Figure 27**).

The prominence of benzimidazole in the field of medicine is exceptional. Etazene (**106**), a benzimidazole opioid that has got strong analgesic activity, is used as a new psychoactive substance (**Figure 27**) [88]. Misuse of certain benzimidazole derivatives can create social crises too.

Tantray and co-workers studied psychiatric disorders like depression and acknowledged the fact that glycogen synthase kinase-3β (GSK-3β) dysfunction is a potential implication [89]. They designed and synthesized several 1,3,4-oxadiazole carboxamides linked to benzimidazoles (**107**) and assessed their *in vitro* GSK-3β inhibition. It was found that these molecules are having antidepressant activity (**Figure 27**).

Hussain et al. synthesized certain benzimidazole analogs (**108**) for the effective management of type-II diabetics [90]. The sulfonamide bearing 2-marcaptobenzimidazoles (**108**), possesses better *in vitro* α-amylase enzyme inhibitory activity while compared with the standard drug, acarbose (**Figure 27**).

Dabigatran is an effective drug having a benzimidazole core as the activity center and is used for the treatment of cardiovascular diseases because of its antithrombin as well as anticoagulant activities. Zhang et al. in 2020 enhanced the activity and bioavailability of dabigatran by adding methyl and methoxy groups into the benzene ring [91]. By studying the anticoagulant action and thrombin inhibition properties of compounds (**109**) in rats, proved the possibility of using these molecules as potential antithrombin drug candidates in the future (**Figure 27**).

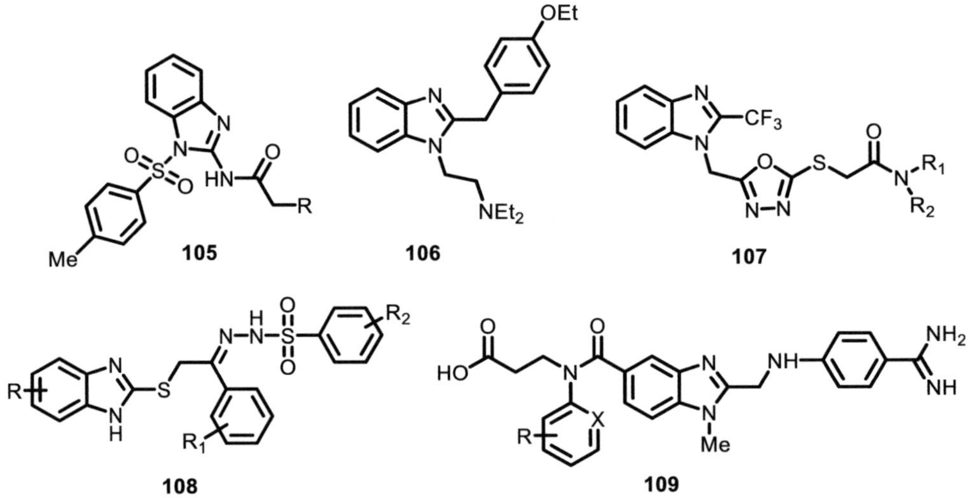

Figure 27.
Benzimidazole derivatives (compound **105, 106, 107, 108** *and* **109***) with antipsychotic, antidiabetic, and anticoagulant activities.*

3. Conclusions

To sum up, benzimidazole is a chemical compound that belongs to the family of heterocyclic aromatic organic compounds. It is a potent biologically important molecule with a noticeable therapeutic activity. Applications of benzimidazole extend to medicinal chemistry. Several advanced research in this area also found out that the aforementioned compound has significant antimicrobial activities especially against many strains of viruses, fungus, bacteria, etc. It is also widely used in medicinal chemistry as an accepted drug against parasites and their allied infections. Benzimidazole is also used as an analgesic and anti-inflammatory agent. Recent studies have also created a lot of attention for the compound since it has an anti-carcinogenic activity like cytotoxicity and hence may become a viable cure for cancer in the future. The applications of benzimidazole cannot be marginalized. It has got a whole spectrum of medicinal agents. Benzimidazole has gained popularity in material science.

Apart from this, the multi-target capability of benzimidazole scaffolds has not been explored extensively. Being a versatile motif, benzimidazole could provide a plethora of novel multi-target ligands against various debilitated pathological conditions. The lack of comprehensive compilation about the SAR of many compounds and the various research reports stemmed the reason for less number of active benzimidazole compounds reaching the market. The existing design of benzimidazole derivatives can be further revised to accommodate potential multitargeting agents, thus enhancing and treating multifactorial disorders. This can be a breakthrough establishment in benzimidazole history.

In short, the importance of this imidazoline compound has been proved by the number of research papers getting published in a short period. This chapter is trying to narrate the formulation as well as execution of benzimidazoles in different fields of medicinal chemistry.

Acknowledgements

The authors would like to acknowledge the support given by the Department of Chemistry, Amrita Vishwa Vidyapeetham, Amritapuri campus for providing the necessary facilities to carry out research work.

Author details

Gopakumar Kavya and Akhil Sivan*
Department of Chemistry, Amrita Vishwa Vidyapeetham, Amritapuri, Kollam, Kerala, India

*Address all correspondence to: akhilsivan@am.amrita.edu

References

[1] Woolley DW. Some biological effects produced by benzimidazole and their reversal by purines. The Journal of Biological Chemistry. 1944;**152**:225-232. DOI: 10.1016/S0021-9258(18)72045-0

[2] Hoebrecker F. Ueber reductionsprodukte der nitracetamidverbindungen. Chemische Berichte. 1872;**5**:920-924. DOI: 10.1002/cber.18720050295

[3] Brink NG, Folkers K. Vitamin B_{12}. VI. 5, 6-dimethylbenzimidazole, a degradation product of vitamin B_{12}. Journal of the American Chemical Society. 1949;**71**:2951. DOI: 10.1021/ja01176a5323

[4] Epstein S. Effect of some benzimidazoles on a vitamin B_{12}-requiring alga. Nature. 1960;**188**: 143-144. DOI: 10.1038/188143a0

[5] Emerson G, Brink NG, Holly FW, Koniuszy F, Heyl D, Folker K. Vitamin B_{12}. VIII. Vitamin B_{12}-like activity of 5, 6-dimethylbenzimidazole and tests on related compounds. Journal of the American Chemical Society. 1950;**72**: 3084-3085. DOI: 10.1021/ja01163a078

[6] Wright JB. Chemistry of benzimidazoles. Chemical Reviews. 1951;**48**:397-541. DOI: 10.1021/cr60151a002

[7] Manna SK, Das T, Samanta S. Polycyclic benzimidazole: Synthesis and photophysical properties. ChemistrySelect. 2019;**4**:8781-8790. DOI: 10.1002/slct.201901941

[8] Spasov AA, Yozhitsa IN, Bugaeva LI, Anisimova VA. Benzimidazole derivatives: Spectrum of pharmacological activity and toxicological properties (a review). Pharmaceutical Chemistry Journal. 1999; **33**:232-243. DOI: 10.1007/BF02510042

[9] Rossignol JF, Maisonneuve H. Benzimidazoles in the treatment of trichuriasis: A review. Annals of Tropical Medicine and Parasitology. 1984;**78**: 135-144. DOI: 10.1080/00034983.1984.11811787

[10] Narasimhan B, Sharma D, Kumar P. Benzimidazole: A medicinally important heterocyclic moiety. Medicinal Chemistry Research. 2012;**21**:269-283. DOI: 10.1007/s00044-010-9533-9

[11] Sivakumar R, Pradeepchandran R, Jayaveera KN, Kumarnallasivan P, Vijaianand PR, Venkatnarayanan R. Benzimidazole: An attractive pharmacophore in medicinal chemistry. International Journal of Pharmaceutical Research. 2011;**3**:19-31

[12] Keri RS, Hiremathad A, Budagumpi S, Nagaraja BM. Comprehensive review in current developments of benzimidazole-based medicinal chemistry. Chemical Biology & Drug Design. 2015;**86**:19-65. DOI: 10.1111/cbdd.12462

[13] Yogita B, Om S. The therapeutic journey of benzimidazoles: A review. Bioorganic & Medicinal Chemistry. 2012;**20**:6208-6236. DOI: 10.1016/j.bmc.2012.09.013

[14] Yadav G, Ganguly S. Structure activity relationship (SAR) study of benzimidazole scaffold for different biological activities: A mini-review. European Journal of Medicinal Chemistry. 2015;**97**:419-443. DOI: 10.1016/j.ejmech.2014.11.053

[15] Boiani M, Gonzalez M. Imidazole and benzimidazole derivatives as

chemotherapeutic agents. Mini Reviews in Medicinal Chemistry. 2005;5:409-424. DOI: 10.2174/1389557053544047

[16] Demirayak Ş, Abu Mohsen U, Çağri Karaburun A. Synthesis and anticancer and anti-HIV testing of some pyrazino [1,2-*a*]benzimidazole derivatives. European Journal of Medicinal Chemistry. 2002;**37**:255-260. DOI: 10.1016/S0223-5234(01)01313-7

[17] Gellis A, Kovacic H, Boufatah N, Vanelle P. Synthesis and cytotoxicity evaluation of some benzimidazole-4,7-diones as bioreductive anticancer agents. European Journal of Medicinal Chemistry. 2008;**43**:1858-1864. DOI: 10.1016/j.ejmech.2007.11.020

[18] Hranjec M, Starčević K, Pavelić SK, Lučin P, Pavelić K, Karminski Zamola G. Synthesis, spectroscopic characterization and antiproliferative evaluation in vitro of novel Schiff bases related to benzimidazoles. European Journal of Medicinal Chemistry. 2011;**46**:2274-2279. DOI: 10.1016/j.ejmech.2011.03.008

[19] Refaat HM. Synthesis and anticancer activity of some novel 2-substituted benzimidazole derivatives. European Journal of Medicinal Chemistry. 2010; **45**:2949-2956. DOI: 10.1016/j.ejmech.2010.03.022

[20] Thimmegowda NR, Nanjunda Swamy S, Ananda Kumar CS, Sunil Kumar YC, Chandrappa S, Yip GW, et al. Synthesis, characterization and evaluation of benzimidazole derivative and its precursors as inhibitors of MDA-MB-231 human breast cancer cell proliferation. Bioorganic and Medicinal Chemistry Letters. 2008;**18**:432-435. DOI: 10.1016/j.bmcl.2007.08.078

[21] Ansari KF, Lal C. Synthesis and evaluation of some new benzimidazole derivatives as potential antimicrobial agents. European Journal of Medicinal Chemistry. 2009;**44**:2294-2299. DOI: 10.1016/j.ejmech.2008.01.022

[22] Kazimierczuk Z, Upcroft JA, Upcroft P, Górska A, Starościak B, Laudy A. Synthesis, antiprotozoal and antibacterial activity of nitro-and halogeno-substituted benzimidazole derivatives. Acta Biochimica Polonica. 2002;**49**:185-195. DOI: 10.18388/abp.2002_3835

[23] Kumar BVS, Vaidya SD, Kumar RV, Bhirud SB, Mane RB. Synthesis and anti-bacterial activity of some novel 2-(6-fluorochroman-2-yl)-1-alkyl/acyl/aroyl-1H-benzimidazoles. European Journal of Medicinal Chemistry. 2006;**41**:599-604. DOI: 10.1016/j.ejmech.2006.01.006

[24] Shutov RV, Kuklina EV, Anisimova NA, EP A'E, Kotova NI, Kuklin VN. Synthesis and biological activity of new pyranochromen-3-carboxamides. Pharmaceutical Chemistry Journal. 2012;**46**:10-14. DOI: 10.1007/s11094-012-0725-0

[25] Köhler P. The biochemical basis of anthelmintic action and resistance. International Journal for Parasitology. 2001;**31**:336-345. DOI: 10.1016/S0020-7519(01)00131-X

[26] Valdez-Padilla D, Rodríguez-Morales S, Hernández-Campos A, Hernández-Luis F, Yépez-Mulia L, Tapia-Contreras A, et al. Synthesis and antiprotozoal activity of novel 1-methylbenzimidazole derivatives. Bioorganic and Medicinal Chemistry. 2009;**17**:1724-1730. DOI: 10.1016/j.bmc.2008.12.059

[27] Gaba M, Singh D, Singh S, Sharma V, Gaba P. Synthesis and pharmacological evaluation of novel 5-substituted-1-(phenylsulfonyl)-2-methylbenzimidazole derivatives as anti-

inflammatory and analgesic agents. European Journal of Medicinal Chemistry. 2010;**45**:2245-2249. DOI: 10.1016/j.ejmech.2010.01.067

[28] Sondhi SM, Rani R, Singh J, Roy P, Agrawal SK, Saxena AK. Solvent free synthesis, anti-inflammatory and anticancer activity evaluation of tricyclic and tetracyclic benzimidazole derivatives. Bioorganic and Medicinal Chemistry Letters. 2010;**20**:2306-2310. DOI: 10.1016/j.bmcl.2010.01.147

[29] Khan SA, Nandan SA. 2-Substituted benzimidazoles as anti-inflammatory and analgesic agents. Indian Journal of Heterocyclic Chemistry. 1997;7:55

[30] Pandey VK, Shukla A. Synthesis and biological activity of isoquinolinyl benzimidazoles. Indian Journal of Chemistry - Section B: Organic and Medicinal Chemistry. 1999;**38**:1381-1383

[31] Yadav LDS, Pal DR. Synthesis and antiviral activity of new 1,3,4-dithi (thiadi)azino(4,5-*a*) benzimidazole derivatives involving oxidative N-S and N-N bond formations. Indian Journal of Chemistry - Section B: Organic and Medcinal Chemistry. 1996;**35B**:748

[32] Starčević K, Kralj M, Ester K, Sabol I, Grce M, Pavelić K. Karminski-Zamola G Synthesis, antiviral and antitumor activity of 2-substituted-5-amidino-benzimidazoles. Bioorganic and Medicinal Chemistry Journal. 2007;**15**: 4419-4426. DOI: 10.1016/j.bmc.2007.04.032

[33] Patil A, Ganguly S, Surana S. A systematic review of benzimidazole derivatives as an antiulcer agent. Rasayan Journal of Chemistry. 2008;**1**: 447-460

[34] McKellar QA, Scott EW. The benzimidazole anthelmintic agents-A review. Journal of Veterinary Pharmacology and Therapeutics. 1990; **13**:223-247. DOI: 10.1111/j.1365-2885.1990.tb00773.x

[35] Noha RM, Abdelhameid MK, Ismail MM, Mohammed MR, Salwa E. Design, synthesis and screening of benzimidazole containing compounds with methoxylated aryl radicals as cytotoxic molecules on (HCT-116) colon cancer cells. European Journal of Medicinal Chemistry. 2021;**209**:112870. DOI: 10.1016/j.ejmech.2020.112870

[36] Anbu S, Paul A, Surendranath K, Solaiman NS, Pombeiro AJL. A benzimidazole-based new fluorogenic differential/sequential chemosensor for Cu^{2+}, Zn^{2+}, CN^{-}, $P_2O_7^{4-}$, DNA, its live-cell imaging and pyrosequencing applications. Sensors Actuators B: Chemical. 2021;**337**:129785. DOI: 10.1016/j.snb.2021.129785

[37] Florio R, Carradori S, Veschi S, Brocco D, Di GT, Cirilli R, et al. Screening of benzimidazole-based anthelmintics and their enantiomers as repurposed drug candidates in cancer therapy. Pharmaceuticals. 2021;**14**:372. DOI: 10.3390/ph14040372

[38] Hsieh C-Y, Ko P-W, Chang Y-J, Kapoor M, Liang Y-C, Chu H-L, et al. Design and synthesis of benzimidazole-chalcone derivatives as potential anticancer agents. Molecules. 2019;**24**: 3259. DOI: 10.3390/molecules24183259

[39] Neverdauskas L, Kaziukonytė P, Brukštus A. Synthesis of benzimidazole derivatives as potential inhibitors for HSP90. Open Readings. 2019;**9**:2019

[40] Hernández-Romero D, Rosete-Luna S, López-Monteon A, Chávez-Piña A, Pérez-Hernández N, Marroquín-Flores J, et al. First-row transition metal compounds containing benzimidazole

ligands: An overview of their anticancer and antitumor activity. Coordination Chemistry Review. 2021;**439**:213930. DOI: 10.1016/j.ccr.2021.213930

[41] Bistrović A, Krstulović L, Harej A, Grbčić P, Sedić M, Koštrun S, et al. Design, synthesis, and biological evaluation of novel benzimidazole amidines as potent multi-target inhibitors for the treatment of non-small cell lung cancer. European Journal of Medicinal Chemistry. 2018;**143**:1616-1634. DOI: 10.1016/j.ejmech.2017.10.061

[42] Akkoç S, Tüzün B, İlhan İÖ, Akkurt M. Investigation of structural, spectral, electronic, and biological properties of 1, 3-disubstituted benzimidazole derivatives. Journal of Molecular Structure. 2020;**1219**:128582. DOI: 10.1016/j.molstruc.2020.128582

[43] Singh I, Luxami V, Paul K. Effective synthesis of benzimidazoles-imidazo [1,2-*a*]pyrazine conjugates: A comparative study of mono-and bis-benzimidazoles for antitumor activity. European Journal of Medicinal Chemistry. 2019;**180**:546-561. DOI: 10.1016/j.ejmech.2019.07.042

[44] Srour AM, Ahmed NS, El-Karim SSA, Anwar MM, El-Hallouty SM. Design, synthesis, biological evaluation, QSAR analysis and molecular modelling of new thiazol-benzimidazoles as EGFR inhibitors. Bioorganic Medicinal Chemistry. 2020;**28**:115657. DOI: 10.1016/j.bmc.2020.115657

[45] Sivaramakarthikeyan R, Iniyaval S, Saravanan V, Lim WM, Mai CW, Ramalingan C. Molecular hybrids integrated with benzimidazole and pyrazole structural motifs: Design, synthesis, biological evaluation, and molecular docking studies. ACS Omega. 2020;**5**:10089-10098. DOI: 10.1021/acsomega.0c00630

[46] Hu M, Yan YL, Zhu B, Chang F, Yu S, Alatan G. A series of Mn(I) photo-activated carbon monoxide-releasing molecules with benzimidazole coligands: Synthesis, structural characterization, CO releasing properties and biological activity evaluation. RSC Advances. 2019; **9**:20505-20512. DOI: 10.1039/c9ra01370a

[47] Prosser KE, Chang SW, Saraci F, Le PH, Walsby CJ. Anticancer copper pyridine benzimidazole complexes: ROS generation, biomolecule interactions, and cytotoxicity. Journal of Inorganic Biochemistry. 2017;**167**:89-99. DOI: 10.1016/j.jinorgbio.2016.11.006

[48] Song W, Xu P, Zhi S, Zhu S, Guo Y, Yang H. Integrated transcriptome and *in vitro* analysis revealed antiproliferative effects on human gastric cancer cells by a benzimidazole-quinoline copper (II) complex. Process Biochemistry. 2021;**102**:286-295. DOI: 10.1016/j.procbio.2021.01.009

[49] Çevik UA, Çavuşoğlu BK, Sağlık BN, Osmaniye D, Levent S, Ilgın S, et al. Synthesis, docking studies and biological activity of new benzimidazole-triazolothiadiazine derivatives as aromatase inhibitor. Molecules. 2020;**25**: 1642. DOI: 10.3390/molecules25071642

[50] Gangrade A, Pathak V, Augelli-Szafran CE, Wei HX, Oliver P, Suto M, et al. Preferential inhibition of Wnt/β-catenin signaling by novel benzimidazole compounds in triple-negative breast cancer. International Journal of Molecular Sciences. 2018;**19**:1524. DOI: 10.3390/ijms19051524

[51] Cheong JE, Zaffagni M, Chung I, Xu Y, Wang Y, Jernigan FE, et al. Synthesis and anticancer activity of novel water soluble benzimidazole carbamates. European Journal of Medicinal Chemistry. 2018;**144**:372-385. DOI: 10.1016/j.ejmech.2017.11.037

[52] Liang Y, Zhou Y, Deng S, Chen T. Microwave-assisted syntheses of benzimidazole-containing selenadiazole derivatives that induce cell-cycle arrest and apoptosis in human breast cancer cells by activation of the ROS/AKT pathway. ChemMedChem. 2016;**11**: 2339-2346. DOI: 10.1002/cmdc.201600261

[53] Husain A, Bhutani M, Parveen S, Khan SA, Ahmad A, Iqbal MA. Synthesis, in vitro cytotoxicity, ADME, and molecular docking studies of benzimidazole-bearing furanone derivatives. Journal of the Chinese Chemical Society. 2021;**68**:362-373. DOI: 10.1002/jccs.202000130

[54] Rashid M. Design, synthesis and ADMET prediction of bis-benzimidazole as anticancer agent. Bioorganic Chemistry. 2020;**96**:103576. DOI: 10.1016/j.bioorg.2020.103576

[55] Shinde VS, Lawande PP, Sontakke VA, Khan A. Synthesis of benzimidazole nucleosides and their anticancer activity. Carbohydrate Research. 2020;**498**:108178. DOI: 10.1016/j.carres.2020.108178

[56] Sireesha R, Sreenivasulu R, Chandrasekhar C, Jadav SS, Pavani Y, Rao MVB, et al. Design, synthesis, anti-cancer evaluation and binding mode studies of benzimidazole/benzoxazole linked β-carboline derivatives. Journal of Molecular Structure. 2021;**1226**:129351. DOI: 10.1016/j.molstruc.2020.129351

[57] Suk FM, Liu CL, Hsu MH, Chuang YT, Wang JP, Liao YJ. Treatment with a new benzimidazole derivative bearing a pyrrolidine side chain overcomes sorafenib resistance in hepatocellular carcinoma. Scientific Reports. 2019;**9**:1-10. DOI: 10.1038/s41598-019-53863-2

[58] Welsh A, Rylands LI, Arion VB, Prince S, Smith GS. Synthesis and antiproliferative activity of benzimidazole-based, trinuclear neutral cyclometallated and cationic, N^N-chelated ruthenium (II) complexes. Dalton Transactions. 2020;**49**:1143-1156. DOI: 10.1039/c9dt03902c

[59] Abdel-Motaal M, Almohawes K, Tantawy MA. Antimicrobial evaluation and docking study of some new substituted benzimidazole-2yl derivatives. Bioinorganic Chemistry. 2020;**101**:103972. DOI: 10.1016/j.bioorg.2020.103972

[60] Ajani OO, Tolu-Bolaji OO, Olorunshola SJ, Zhao Y, Aderohunmu DV. Structure-based design of functionalized 2-substituted and 1, 2-disubstituted benzimidazole derivatives and their *in vitro* antibacterial efficacy. Journal of Advanced Research. 2017;**8**:703-712. DOI: 10.1016/j.jare.2017.09.003

[61] Bistrović A, Krstulović L, Stolić I, Drenjančević D, Talapko J, Taylor MC, et al. Synthesis, anti-bacterial and anti-protozoal activities of amidinobenzimidazole derivatives and their interactions with DNA and RNA. Journal of Enzyme Inhibition and Medicinal Chemistry. 2018;**33**:1323-1334. DOI: 10.1080/14756366.2018.1484733

[62] Bhavsar ZA, Acharya PT, Jethava DJ, Patel DB, Vasava MS, Rajani DP, et al. Microwave assisted synthesis, biological activities, and *in silico* investigation of some benzimidazole derivatives. Journal of Heterocyclic Chemistry. 2020;**57**: 4215-4238. DOI: 10.1002/jhet.4129

[63] Jiménez-Juárez R, Cruz-Chávez W, de Jesús-Ramírez N, Castro-Ramírez GI, Uribe-González I, Martínez-Mejía G, et al. Synthesis and antimycobacterial activity of 2,5-disubstituted and 1,2,5-

trisubstituted benzimidazoles. Frontiers in Chemistry. 2020;**8**:1-12. DOI: 10.3389/fchem.2020.00433

[64] Penieres-Carrillo JG, Ríos-Guerra H, Pérez-Flores J, Rodríguez-Molina B, Torres-Reyes Á, Barrera-Téllez F, et al. Reevaluating the synthesis of 2,5-disubstituted-1*H*-benzimidazole derivatives by different green activation techniques and their biological activity as antifungal and antimicrobial inhibitor. Journal of Heterocyclic Chemistry. 2020; **57**:436-455. DOI: 10.1002/jhet.3801

[65] Khan S, Kale M, Siddiqui F, Nema N. Novel pyrimidine-benzimidazole hybrids with antibacterial and antifungal properties and potential inhibition of SARS-CoV-2 main protease and spike glycoprotein. Digital Chinese Medicine. 2021;**4**:102-119. DOI: 10.1016/j.dcmed.2021.06.004

[66] Zha GF, Preetham HD, Rangappa S, Sharath Kumar KS, Girish YR, Rakesh KP, et al. Benzimidazole analogues as efficient arsenals in war against methicillin-resistance staphylococcus aureus (MRSA) and its SAR studies. Bioorganic Chemistry. 2021;**115**:105175. DOI: 10.1016/j.bioorg.2021.105175

[67] Marinescu M. Synthesis of antimicrobial benzimidazole–pyrazole compounds and their biological activities. Antibiotics. 2021;**10**:1002. DOI: 10.1016/j.molstruc.2016.10.066

[68] Karaburun AÇ, Kaya Çavuşoğlu B, Acar Çevik U, Osmaniye D, Sağlık BN, Levent S, et al. Synthesis and antifungal potential of some novel benzimidazole-1,3,4-oxadiazole compounds. Molecules. 2019;**24**:191. DOI: 10.3390/molecules24010191

[69] Aroso RT, Guedes RC, Pereira MM. Synthesis of computationally designed 2,5(6)-benzimidazole derivatives *via* Pd-Catalyzed reactions for potential E. coli DNA gyrase B inhibition. Molecules. 2021;**26**:1326. DOI: 10.3390/molecules26051326

[70] Chen M, Su S, Zhou Q, Tang X, Liu T, Peng F, et al. Antibacterial and antiviral activities and action mechanism of flavonoid derivatives with a benzimidazole moiety. Journal of Saudi Chemical Society. 2021;**25**:101194. DOI: 10.1016/j.jscs.2020.101194

[71] Karaca Gençer H, Acar Çevik U, Levent S, Sağlık BN, Korkut B, Özkay Y, et al. New benzimidazole-1, 2, 4-triazole hybrid compounds: Synthesis, anticandidal activity and cytotoxicity evaluation. Molecules. 2017;**22**:507. DOI: 10.3390/molecules22040507

[72] Gullapelli K, Brahmeshwari G, Ravichander M, Kusuma U. Synthesis, antibacterial and molecular docking studies of new benzimidazole derivatives. Egyptan Journal of Basic Applied Sciences. 2017;**4**:303-309. DOI: 10.1016/j.ejbas.2017.09.002

[73] Wang YN, Bheemanaboina RRY, Gao WW, Kang J, Cai GX, Zhou CH. Discovery of benzimidazole–quinolone hybrids as new cleaving agents toward drug-resistant pseudomonas aeruginosa DNA. ChemMedChem. 2018;**13**: 1004-1017. DOI: 10.1002/cmdc.201700739

[74] Marinescu M, Tudorache DG, Marton GI, Zalaru CM, Popa M, Chifiriuc MC, et al. Density functional theory molecular modeling, chemical synthesis, and antimicrobial behaviour of selected benzimidazole derivatives. Journal of Molecular Structure. 2017;**1130**: 463-471. DOI: 10.1016/j.molstruc.2016.10.066

[75] Abraham R, Prakash P, Mahendran K, Ramanathan M. A novel

series of N-acyl substituted indole-linked benzimidazoles and naphthoimidazoles as potential antiinflammatory, antibiofilm and antimicrobial agents. Microbial Pathogenesis. 2018;**114**: 409-413. DOI: 10.1016/j.micpath.2017.12.021

[76] Antoci V, Cucu D, Zbancioc G, Moldoveanu C, Mangalagiu V, Amariucai-Mantu D, et al. Bis-(imidazole/benzimidazole)-pyridine derivatives: Synthesis, structure and antimycobacterial activity. Future Medicinal Chemistry. 2020; **12**:207-222. DOI: 10.4155/fmc-2019-0063

[77] Ersan RH, Yuksel A, Ertan-Bolelli T, Dogen A, Burmaoglu S, Algul O. One-pot synthesis of novel benzimidazoles with a naphthalene moiety as antimicrobial agents and molecular docking studies. Journal of the Chinese Chemical Society. 2021;**68**:374-383. DOI: 10.1002/jccs.202000125

[78] Sirim MM, Krishna VS, Sriram D, Tan OU. Novel benzimidazole-acrylonitrile hybrids and their derivatives: Design, synthesis and antimycobacterial activity. European Journal of Medicinal Chemistry. 2020; **188**:112010. DOI: 10.1016/j.ejmech.2019.112010

[79] Taman A, El-Bardicy S, Tadros M, Ayoub M, Mansour B, El-Shehabi F, et al. In vitro efficacy of new synthetic benzimidazole-related compounds against Schistosoma mansoni adult worms. Asian Pacific Journal of Tropical Medicine. 2020;**13**:566-572. DOI: 10.4103/1995-7645.296725

[80] Mavrova A, Anichina K, Izevbekhai O, Vutchev D, Popova-daskalova G, Yancheva D, et al. New 1, 3-disubsituted benzimidazol-2-ones as a promising scaffold for the antitrichinellosis agents development. Journal of Chemical Technology and Metallurgy. 2021;**56**:3-9

[81] Pauli I, Ferreira LG, de Souza ML, Oliva G, Ferreira RS, Dessoy MA, et al. Molecular modeling, and structure-activity relationships for a series of benzimidazole derivatives as cruzain inhibitors. Future Medicinal Chemistry. 2017;**9**:641-657. DOI: 10.4155/fmc-2016-0236

[82] Tonelli M, Gabriele E, Piazza F, Basilico N, Parapini S, Tasso B, et al. Benzimidazole derivatives endowed with potent antileishmanial activity. Journal of Enzyme Inhibition and Medicinal Chemistry. 2018;**33**: 210-226. DOI: 10.1080/14756366.2017.1410480

[83] Türkan F. Investigation of the toxicological and inhibitory effects of some benzimidazole agents on acetylcholinesterase and butyrylcholinesterase enzymes. Archives of Physiology and Biochemistry. 2021; **127**:97-101. DOI: 10.1080/13813455.2019.1618341

[84] Bessières M, Plebanek E, Chatterjee P, Shrivastava-Ranjan P, Flint M, Spiropoulou CF, et al. Design, synthesis and biological evaluation of 2-substituted-6-[(4-substituted-1-piperidyl)methyl]-1*H*-benzimidazoles as inhibitors of ebola virus infection. European Journal of Medicinal Chemistry. 2021;**214**:113211. DOI: 10.1016/j.ejmech.2021.113211

[85] Ibba R, Carta A, Madeddu S, Caria P, Serreli G, Piras S, et al. Inhibition of enterovirus A71 by a novel 2-phenyl-benzimidazole derivative. Viruses. 2021;**13**: 58. DOI: 10.3390/v13010058

[86] Wada Y, Orba Y, Sasaki M, Kobayashi S, Carr MJ, Nobori H, et al. Discovery of a novel antiviral agent

targeting the nonstructural protein 4 (nsP4) of chikungunya virus. Virology. 2017;**505**:102-112. DOI: 10.1016/j.virol.2017.02.014

[87] Imran M, Shah FA, Nadeem H, Zeb A, Faheem M, Naz S, et al. Synthesis and biological evaluation of benzimidazole derivatives as potential neuroprotective agents in an ethanol-induced rodent model. ACS Chemical Neuroscience. 2021;**12**:489-505. DOI: 10.1021/acschemneuro.0c00659

[88] Siczek M, Zawadzki M, Siczek M, Chłopaś-Konowałek A, Szpot P. Etazene (N, N-diethyl-2-{[(4-ethoxyphenyl) methyl]-1*H*-benzimidazol-1-yl}-ethan-1-amine (dihydrochloride)): A novel benzimidazole opioid NPS identified in seized material: Crystal structure and spectroscopic characterization. Forensic Toxicology. 2021;**39**:146-155. DOI: 10.1007/s11419-020-00552-9

[89] Tantray MA, Khan I, Hamid H, Alam MS, Dhulap A, Kalam A. Synthesis of benzimidazole-linked-1,3,4-oxadiazole carboxamides as GSK-3β inhibitors with *in vivo* antidepressant activity. Bioorganic Chemistry. 2018;**77**: 393-401. DOI: 10.1016/j.bioorg.2018.01.040

[90] Hussain S, Taha M, Rahim F, Hayat S, Zaman K, Iqbal N, et al. Synthesis of benzimidazole derivatives as potent inhibitors for α-amylase and their molecular docking study in management of type-II diabetes. Journal of Molecular Structure. 2021;**1232**: 130029. DOI: 10.1016/j.molstruc.2021.130029

[91] Zhang T, Liu Q, Ren Y. Design, synthesis and biological activity evaluation of novel methyl substituted benzimidazole derivatives. Tetrahedron. 2020;**76**:131027. DOI: 10.1016/j.tet.2020.131027

Chapter 5

Synthesis, Characterization and Antimicrobial Properties of Novel Benzimidazole Amide Derivatives Bearing Thiophene Moiety

Vinayak Adimule, Pravin Kendrekar and Sheetal Batakurki

Abstract

In the present investigation, novel amide derivatives of benzimidazole (**4a-f**) with different thiophene acids (**a-f**) coupled in the presence of 1-[Bis (dimethylamino) methylene]-1H-1, 2, 3-triazolo [4, 5-b] pyridinium 3-oxide hexafluorophosphate (HATU) reagent at room temperature and as-synthesized derivatives were characterized by (^{1}H-NMR and ^{13}C-NMR) proton and carbon magnetic resonance, and high-performance liquid chromatography (HPLC) analytical techniques. The amide derivatives were tested for *in vitro* antimicrobial and antifungal activity and ciprofloxacin was used as standard. The antifungal activity was tested with *Carbendazim* and *Fenbendazole* cell lines using *clotrimazole* standard drug. The results indicated the potential activity toward *S. bacillus* with compounds having IC $_{50}$ of **4 (a)**, **4 (b)**, **4 (d)** and **4 (e)** against antimicrobial strains with IC_{50} of **51.8** μm, **57.4** μm, **54.5** μm and **56.5** μm respectively. However, compounds **4 (a)**, **4 (c)** and **4 (d)** showed greater inhibitions against *Carbendazim* fungal cell line with IC_{50} of **22.9**, **26.8** and **28.8** μm. On the other hand IC_{50} values of the *Fenbendazole* for compounds **4 (a)**, **4(c)** and **4 (d)** were found to be **12.7**, **10.2** and **12.7** μm, respectively. The thiophene-substituted benzimidazole amide derivatives are the potential candidate drug for antibacterial and antifungal activity.

Keywords: thiophene, antibacterial, antifungal, benzimidazole, amides, HATU

1. Introduction

In the present investigation, organic and medicinal chemistry involves the nitrogen containing heterocycles especially Benzimidazole derivatives [1, 2]. Benzimidazole has become nature and synthetic active structural part in the biological characteristics such as antibacterial [3], anticancer [4], antiviral [5], antifungal [6] and antioxidant [7] etc. It is widely accepted that the amidino moieties at the benzimidazole substituents terminal would generate numerous pathophysiological and biochemical processes in the human body. Additional biological activity can be found with the benzimidazole substituents carrying amide moiety and tetracyclic derivatives interaction with DNA is large enough to encounter the selectivity towards the drug molecules. Benzimidazole is indispensable nucleus for the discovery of the new biologically important molecules. Literature reports suggest that benzimidazole has potential interest in antimicrobial [8, 9] and anticancer agents [10, 11]. New class of oxadiazole and thiadiazoles containing thiophene and phenyl substituents reported for enhanced anticancer activity [12–15]. Benzimidazole nucleus structurally analogues to purine and its derivatives and exhibit the synthesis of nucleic acids. Several DNA molecules will be cleaved upon interaction with the benzimidazole moiety and inhibit the growth of the microbial strains [16–18]. The development of new antimicrobial agents remains on priority [19]. Furthermore, Bisbezimidazole derivatives found to be biologically active towards antibacterial [20], antimicrobial [21] and anticancer [22] activities. On the other hand, benzylvanilline benzimidazole [23] derivatives were found to cleave DNA and potent towards leukemic cell lines. Distance between the benzimidazole molecule and ester containing phenyl group is very important factor for the antifungal activity. Extending the spacer between the benzimidazole and ester or amide substituents become important factor for the increased antifungal activity. Antitumor activity depends on the distance between the benzimidazole and spacer link of the ester or amide molecules. Thiophene and their derivatives which are sulfur compounds widely studied for their antifungal activity [24–28]. Some of the thiophene containing derivatives such as thicyofen, ethaboxam, silthiopham and penthiopyrad were widely used in agriculture as antifungal compounds [29–30]. Recently certain amide, thiazole, 1, 3, 4-oxadiazoles has been designed, synthesized and studied for their enhanced anticancer properties [31–36]. Incorporation of the thiophene moiety would enhance the antifungal activity of the benzimdiazole derivatives.

Based on the above literature evidences, Author envisaged by constructing the novel benzimidazole containing thiophene derivative could increase the antimicrobial and antifungal activity of the synthesized compounds. All the synthesized derivatives were characterized by ^{1}H-NMR, ^{13}C-NMR, LCMS spectroscopic studies and tested against microbial and fungal cell lines. The results indicated potential activity towards *S. bacillus* tested compound with IC_{50} of **4 (a)**, **4 (b)**, **4 (d)** and **4 (e)** derivatives against antimicrobial strains with IC_{50} of **51.8** μm, **57.4** μm, **54.5** μm and **56.5** μm respectively. However, compounds **4 (a)**, **4 (c)** and **4 (d)** showed greater inhibitions against *Carbendazim* fungal cell line with IC_{50} of **22.9** μm **26.8** μm and **28.8** μm respectively. On the other hand IC_{50} values of the *Fenbendazole* for compounds **4 (a)**, **4(c)** and **4 (d)** was found to be **12.7** μm, **10.2** μm and **12.7** μm respectively.

2. Materials and methods

All the chemicals, reagents and solvents are procured from Sigma-Aldrich, S-d fine chemicals Ltd., Spectrochem Ltd. and used without any further purification. Intermediate chemicals purchased directly (99.8% purity, thiophene-2-carboxylic acid, Sigma Aldrich India (a)), (98% purity, (thiophen-3-yl) acetic acid Sigma Aldrich India (b)), (99.1% purity, 3,6-dichloro-1-benzothiophene-2-carboxylic acid, Matrix Scientific, India (c)), (99.2% purity, 5-(4-fluorophenyl)thiophene-2-carboxylic acid.

Sigma Aldrich, India (d)), (98% purity, 4-bromo-5-chlorothiophene-2-carboxylic acid, Chem Src, China (e)), (99.2% purity, 3-chloro-1-benzothiophene-2-carboxylic acid, Sigma-Aldrich India (f)) and used without any further purifications. ^{1}H-NMR, ^{13}C-NMR spectral analysis was carried out using 300 MHz-1.2 GHz consisting of cryostat with excellent thermal efficiency, available with high performance vibration isolator make from Bruker Ascend series. LCMS (liquid crystal mass spectrometry)

Triple quadrapole series of GCMS-TQ8050 NX fitted with high efficient trace level detection. In order to prepare the culture medium, the synthesized compounds were dissolved in DMSO and diluted with culture broth solution to 1 mg/mL. Serial dilutions were made to reach up to the 10 mL of the final concentration. About 100 μL of the each of the solution were distributed to 96 well plates and the sterility control and growth control solutions were placed into it. About 5 μL of the test and the growth solutions were inoculated into the well plates. All the experiments were carried out in triplicate. Bacterial growth was detected former by optical density (ELISA reader, CLX800 Biotech Instruments) and after by addition of 20 μL of an INT alcoholic solution (0.5 mg/mL). 7 mm filter paper discs (Whatman, no. 3) were impregnated with 10 mL of each of the different dilutions [37, 38]. The discs were allowed to remain at RT until complete diluent evaporation and kept under refrigeration until ready to be used. Discs loaded with synthesized derivatives which were placed onto the surface of the agar. Commercial chloramphenicol discs and paper discs impregnated with 20 mL of diluents used to dilute concentration of the synthesized products were used as control.

3. Experimental

3.1 Synthesis of 1*H*-benzimidazole-2-carboxylic acid

In a 100 mL round bottom flask fitted with a reflux condenser, 2-methyl-1*H*-benzimidazole (5.69 g, 1.1 mole), $KMnO_4$ (4 mole), K_2CO_3 (3 mole) were dissolved in 100 mL of absolute ethanol and 20 mL of water under stirring. The contents of the reaction mixture heated to 90 C for 5 h. The progress of the reaction was monitored by TLC (thin-layer chromatography) (ethyl acetate: hexane: 40:60). After completion of the reaction, contents of the RB flask cooled to room temperature (RT), filtered over celite bed, filtrate was concentrated to half of its initial volume, acidified with concentrated HCl to $P^H \sim 2$. Off white colored solid separated out was filtered, washed with cold water and dried. Yield: 5.2 g. The solid was taken to next step without any purification [39, 40].

3.2 Synthesis of 1-(1*H*-benzimidazol-2-yl) methanamine

In a 100 mL RB flask fitted with an inert nitrogen atmosphere, 1H-benzimidazole-2-carboxylic acid (5.2 g) dissolved in dry THF solvent and $LiAlH_4$ (1 mole) were added under stirring at RT. Progress of the reaction was monitored by TLC, after completion of the reaction. $LiAlH_4$ was quenched in carbonate solution. The reaction mixture was concentrated to remove the organic solvent and filtered. The product was extracted in ethyl acetate (100 mL × 3 times), washed with Na_2CO_3 (25 mL × 2 times), brine (25 mL × 2 times) and dried over Na_2SO_4. Solid that are separated out was filtered in 10 mL of n-hexane solvent and obtained as 1-(1*H*-benzimidazol-2-yl) methanamine (**Figure 1**). Yield: 3.87 g.

3.3 General procedure for the synthesis of benzimidazole amide derivatives

In a 100 mL RB flask fitted with a condenser, 1-(1*H*-benzimidazol-2-yl) methanamine (1mole) added with different thiophene substituted acids (**a-f**), Dichloromethane (DCC) (10 mL), HATU (0.25mole), Triethyl amine (TEA) (1.5mole) and stirred at RT for 24 h, after completion of the reaction (monitored by

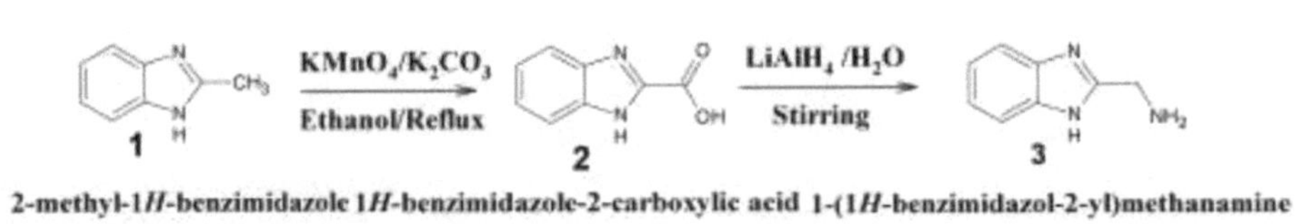

Figure 1.
Synthetic pathway of Benzimidazole intermediates.

TLC), solvent and other volatile reagents are evaporated, solid obtained was mixed with 60–120 silica gel and purified by column-chromatography (ethyl acetate: hexane: 30:70), Solids obtained after the solvent evaporation was used for the antimicrobial property studies [41].

3.4 Characterization of the synthesized compounds (Benzimidazole 2, 3 and 4a-4f)

^{1}H-NMR Spectrum of the compounds 2, 3, 4a to 4 f.

^{1}H-NMR, ^{13}C-NMR and LCMS analysis of the synthesized compounds.

Compound 2.

Yellow solid; m.p-157.8°C; Yield = 97.8%; LCMS-98.1%; ^{1}H-NMR: δ ^{1}H NMR: δ 7.04 (1H, ddd, J = 7.7, 6.9, 1.3 Hz), 7.23 (1H, ddd, J = 8.2, 6.9, 1.7 Hz), 7.77 (1H, ddd, J = 8.2, 1.3, 0.5 Hz), 7.90 (1H, ddd, J = 7.7, 1.7, 0.5 Hz). ^{13}C-NMR: δ ^{13}C NMR: δ 114.3 (1C, s), 118.4 (1C, s), 128.1–128.3 (2C, 128.2 (s), 128.2 (s), 137.9 (1C, s), 138.4 (1C, s), 150.9 (1C, s), 156.7 (1C, s) (**Figure 2**) (**Figures 3–6**).

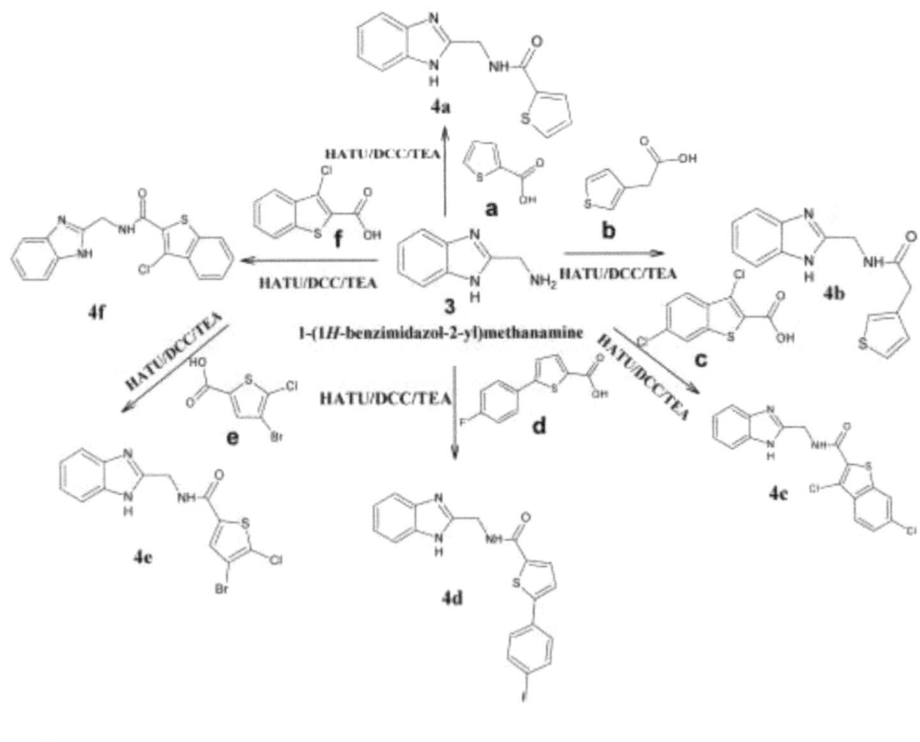

Figure 2.
Synthesis of amide derivatives of Benzimidazoles.

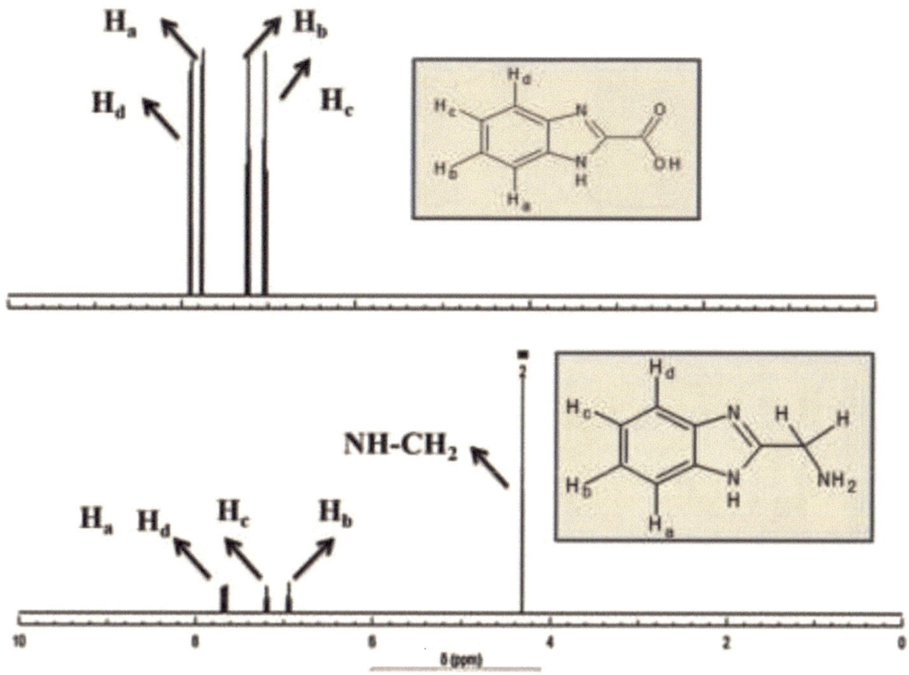

Figure 3.
^{1}H-NMR of the spectrum of 2 and 3.

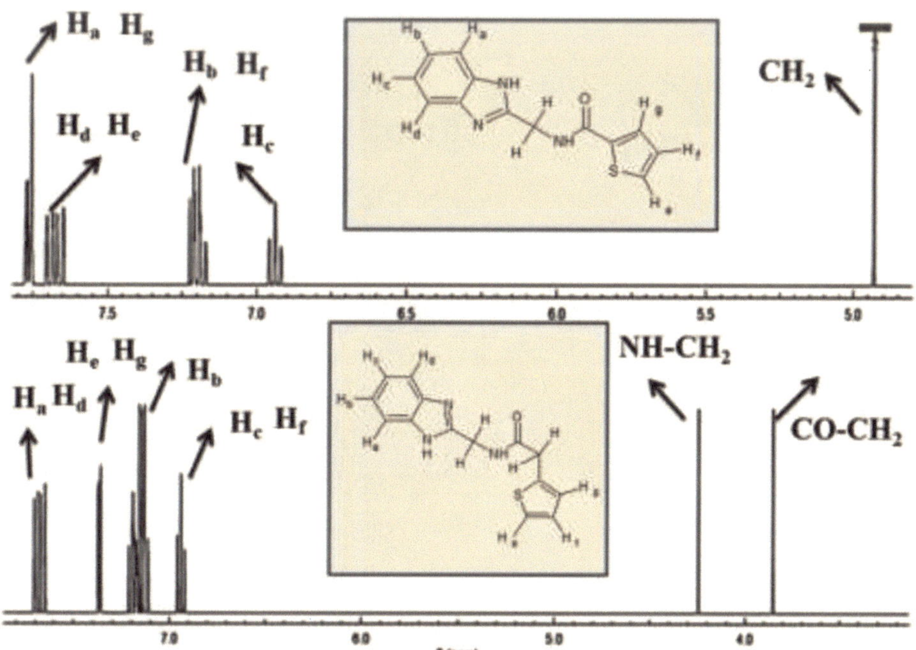

Figure 4.
^{1}H-NMR of the spectrum 4a and 4b.

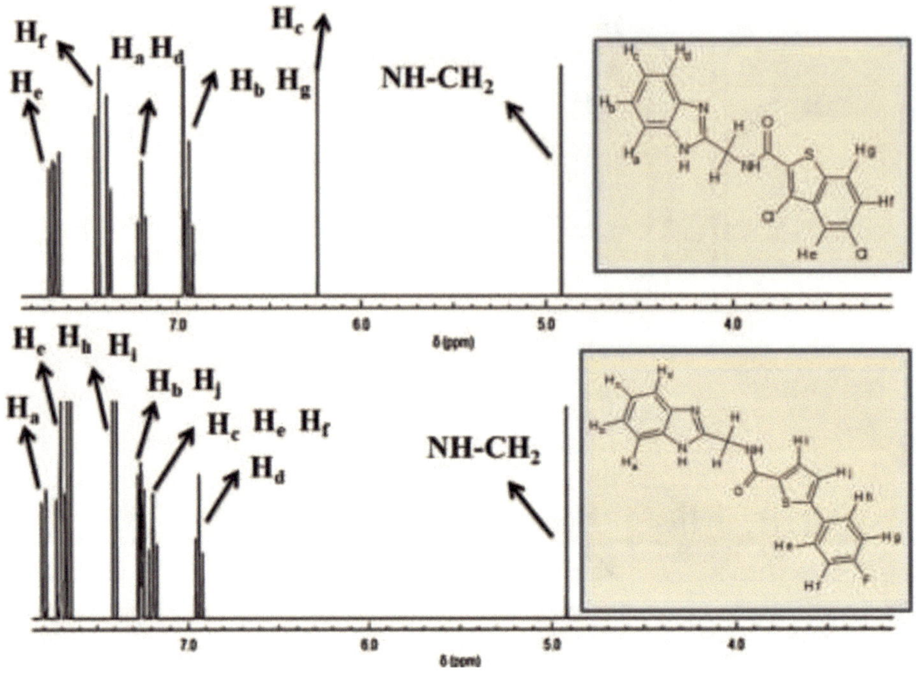

Figure 5.
^{1}H-NMR of the compounds 4c and 4d.

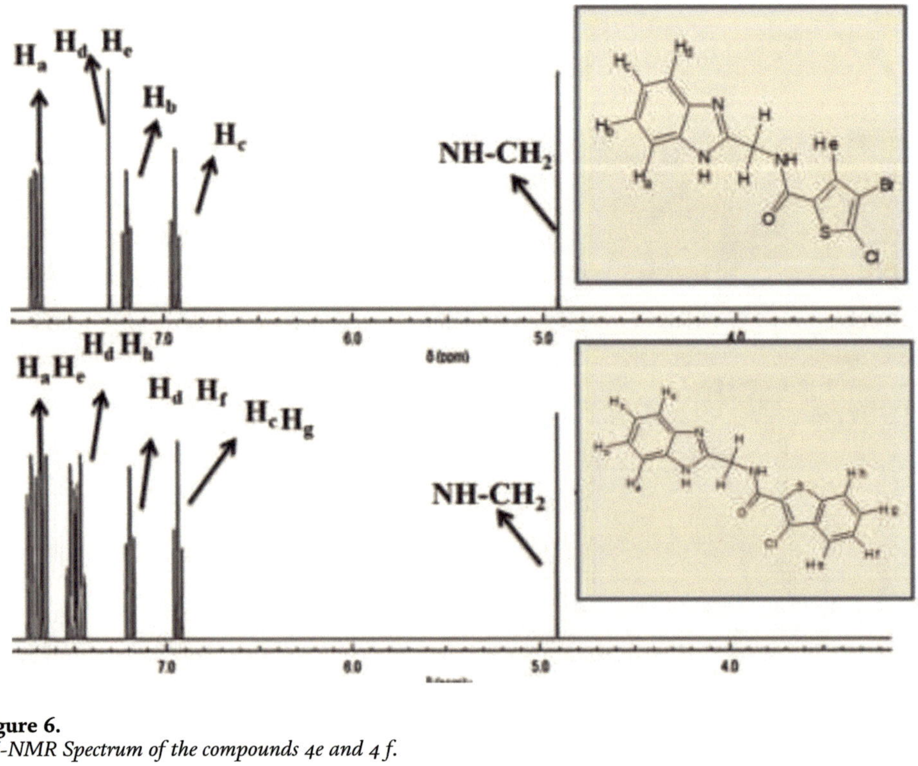

Figure 6.
^{1}H-NMR Spectrum of the compounds 4e and 4 f.

Compound 3.

Pale yellow solid; m.p-159.2°C; Yield = 98.8%; LCMS-98.7%;^{1}H-NMR: δ ^{1}H NMR: *δ* 4.32 (2H, s), 6.94 (1H, ddd, *J* = 7.7, 7.6, 1.2 Hz), 7.19 (1H, ddd, *J* = 8.1, 7.6, 1.4 Hz), 7.59–7.75 (2H, 7.65 (ddd, *J* = 8.1, 1.2, 0.5 Hz), 7.69 (ddd, *J* = 7.7, 1.4, 0.5 Hz). ^{13}C-NMR: *δ* ^{13}C NMR: δ 49.1 (1C, s), 114.3 (1C, s), 118.4 (1C, s), 128.1–128.3 (2C, 128.2 (s), 128.2 (s), 137.9 (1C, s), 138.4 (1C, s), 150.9 (1C, s).

Compound 4a.

Pale yellow solid; m.p-163.4°C; Yield = 97.8%; LCMS- 98.1%;^{1}H NMR: *δ* 4.93 (2H, s), 6.94 (1H, ddd, *J* = 7.7, 7.6, 1.2 Hz), 7.12–7.27 (2H, 7.19 (ddd, *J* = 8.1, 7.6, 1.4 Hz), 7.21 (dd, *J* = 7.2, 5.0 Hz)), 7.60–7.83 (4H, 7.66 (ddd, *J* = 8.1, 1.2, 0.5 Hz), 7.69 (ddd, *J* = 7.7, 1.4, 0.5 Hz), 7.76 (dd, *J* = 5.0, 1.2 Hz), 7.76 (dd, *J* = 7.2, 1.2 Hz). ^{13}C-NMR: *δ* ^{13}C NMR: *δ* 44.7 (1C, s), 114.3 (1C, s), 118.4 (1C, s), 127.5 (1C, s), 127.8 (1C, s), 128.1–128.3 (2C, 128.2 (s), 128.2 (s), 131.0 (1C, s), 137.9 (1C, s), 138.4 (1C, s), 139.9 (1C, s), 150.9 (1C, s), 160.1 (1C, s).

Compound 4b.

Off white colored solid; m.p-171.8°C; Yield = 97.8%; LCMS- 98.3%; ^{1}H NMR: *δ* 3.88 (2H, s), 4.99 (2H, s), 6.94 (1H, ddd, *J* = 7.7, 7.6, 1.2 Hz), 7.10–7.26 (2H, 7.16 (dd, *J* = 7.5, 5.0 Hz), 7.19 (ddd, *J* = 8.1, 7.6, 1.4 Hz), 7.31–7.45 (2H, 7.37 (dd, *J* = 5.0, 1.3 Hz), 7.39 (dd, *J* = 7.5, 1.3 Hz), 7.59–7.76 (2H, 7.66 (ddd, *J* = 8.1, 1.2, 0.5 Hz), 7.69 (ddd, *J* = 7.7, 1.4, 0.5 Hz). ^{13}C-NMR: *δ* ^{13}C NMR: *δ* 29.2 (1C, s), 44.7 (1C, s), 114.3 (1C, s), 118.4 (1C, s), 126.6 (1C, s), 127.5 (1C, s), 127.8 (1C, s), 128.1–128.3 (2C, 128.2 (s), 128.2 (s), 137.9 (1C, s), 138.1 (1C, s), 138.4 (1C, s), 150.9 (1C, s), 172.7 (1C, s).

Compound 4c.

Pale Yellow Solid; m.p^{-174}.5°C; Yield = 99.4%; LCMS- 97.8%; ^{1}H NMR: *δ* 4.90 (2H, s), 6.94 (1H, ddd, *J* = 7.7, 7.6, 1.2 Hz), 7.19 (1H, ddd, *J* = 8.1, 7.6, 1.4 Hz), 7.42 (1H, dd, *J* = 8.6, 1.9 Hz), 7.60–7.76 (2H, 7.66 (ddd, *J* = 8.1, 1.2, 0.5 Hz), 7.69 (ddd, *J* = 7.7, 1.4, 0.5 Hz), 7.94–8.07 (2H, 8.00 (dd, *J* = 1.9, 0.5 Hz), 8.01 (dd, *J* = 8.6, 0.5 Hz)). ^{13}C-NMR: *δ* ^{13}C NMR: *δ* 44.7 (1C, s), 114.3 (1C, s), 118.4–118.5 (2C, 118.4 (s), 118.5 (s)), 127.0 (1C, s), 127.8 (1C, s), 128.1–128.3 (2C, 128.2 (s), 128.2 (s), 128.7 (1C, s), 130.3 (1C, s), 132.6 (1C, s), 136.6 (1C, s), 137.9 (1C, s), 138.4 (1C, s), 139.3 (1C, s), 150.9 (1C, s), 160.1 (1C, s).

Compound 4d.

Yellow solid; m.p-157.8°C; Yield = 97.8%; LCMS- 96.8%; ^{1}H NMR: *δ* 4.92 (2H, s), 6.94 (1H, ddd, *J* = 7.7, 7.6, 1.2 Hz), 7.12–7.32 (3H, 7.19 (ddd, *J* = 8.1, 7.6, 1.4 Hz), 7.25 (ddd, *J* = 8.9, 1.4, 0.5 Hz)), 7.40 (1H, d, *J* = 8.7 Hz), 7.60–7.82 (5H, 7.66 (ddd, *J* = 8.1, 1.2, 0.5 Hz), 7.66 (d, *J* = 8.7 Hz), 7.69 (ddd, *J* = 7.7, 1.4, 0.5 Hz), 7.76 (ddd, *J* = 8.9, 1.5, 0.5 Hz). ^{13}C-NMR: *δ* ^{13}C NMR: δ 44.7 (1C, s), 114.3 (1C, s), 115.4 (2C, s), 118.4 (1C, s), 124.0 (1C, s), 127.8 (2C, s), 128.1–128.3 (2C, 128.2 (s), 128.2 (s)), 129.6 (1C, s), 134.3 (1C, s), 137.8–138.0 (2C, 137.8 (s), 137.9 (s), 138.4 (1C, s), 150.9 (1C, s), 151.1 (1C, s), 160.1 (1C, s), 162.5 (1C, s).

Compound 4e.

Pale Yellow Solid; m.p-179.2°C; Yield = 95.1%; LCMS- 98.2%; ^{1}H NMR: *δ* 4.92 (2H, s), 6.94 (1H, ddd, *J* = 7.7, 7.6, 1.2 Hz), 7.19 (1H, ddd, *J* = 8.1, 7.6, 1.4 Hz), 7.60–7.76 (3H, 7.66 (ddd, *J* = 8.1, 1.2, 0.5 Hz), 7.65 (s), 7.69 (ddd, *J* = 7.7, 1.4, 0.5 Hz). ^{13}C-NMR: *δ* ^{13}C NMR: *δ* 44.7 (1C, s), 110.5 (1C, s), 114.3 (1C, s), 118.4 (1C, s), 123.0 (1C, s), 128.0 (1C, s), 128.1–128.3 (2C, 128.2 (s), 128.2 (s), 134.4 (1C, s), 137.9 (1C, s), 138.4 (1C, s), 150.9 (1C, s), 160.1 (1C, s).

Compound 4f.

White colored solid; m.p-178.5°C; Yield = 94.3%; LCMS- 99.2%; ^{1}H NMR: *δ* 4.91 (2H, s), 6.94 (1H, ddd, *J* = 7.7, 7.6, 1.2 Hz), 7.19 (1H, ddd, *J* = 8.1, 7.6, 1.4 Hz), 7.40–7.59

(2H, 7.47 (ddd, *J* = 8.0, 7.8, 1.4 Hz), 7.52 (ddd, *J* = 8.5, 7.8, 1.6 Hz), 7.60–7.80 (3H, 7.66 (ddd, *J* = 8.1, 1.2, 0.5 Hz), 7.69 (ddd, *J* = 7.7, 1.4, 0.5 Hz), 7.74 (ddd, *J* = 8.0, 1.6, 0.5 Hz)), 8.47 (1H, ddd, *J* = 8.5, 1.4, 0.5 Hz). ^{13}C-NMR: δ ^{13}C NMR: δ 44.7 (1C, s), 114.3 (1C, s), 118.4–118.5 (2C, 118.4 (s), 118.5 (s), 122.5 (1C, s), 123.3 (1C, s), 128.1–128.3 (2C, 128.2 (s), 128.2 (s), 128.3–128.6 (2C, 128.4 (s), 128.5 (s), 132.6 (1C, s), 136.1 (1C, s), 136.6 (1C, s), 137.9 (1C, s), 138.4 (1C, s), 150.9 (1C, s), 160.1 (1C, s).

4. Results and discussion

4.1 Antimicrobial activity of the synthesized derivatives

Antibacterial and antifungal activities of the synthesized compounds were tested by Agar well diffusion method. One mL of the fresh bacterial or fungi was placed in the sterile petri dish. Cooled Muller Hi was placed over it and upon solidification. 5 μL of the compounds dissolved in DMSO solvent was added to respective wells. The concentration has been fixed as per the previous reported literature [42, 43]. The Agar well plates were incubated for 30 minutes and subsequently kept at 35°C for 24 h. Antimicrobial activity was detected by measuring the zone of inhibition (including the wells diameter) appeared after the incubation period. DMSO at a concentration of 10% was employed as a negative control. All tested samples and their antimicrobial activity were expressed in terms of minimum inhibitory concentration (MIC values) (**Tables 1–3**). Different concentrations of the test solution are prepared and zone of inhibition and MIC values compared with the standards used for the evaluation [44]. The MIC was considered as the lowest concentration which inhibited the growth of the respective microorganisms. All assays were performed in triplicate. DMSO was served as a control for all the synthesized samples. **Tables 1** and **2** depicts the minimum inhibitory concentration of the antibacterial and antifungal activity (**Table 3**) of the synthesized derivatives. The results indicated potential activity towards *S. bacillus* tested compound with IC_{50} of **4 (a)**, **4 (b)**, **4 (d)** and 4 **(e)** derivatives against antimicrobial strains with IC_{50} of **51.8** μm, **57.4** μm, **54.5** μm and **56.5** μm respectively. However, compounds **4 (a)**, **4 (c)** and **4 (d)** showed greater inhibitions against *Carbendazim* fungal cell line with IC_{50} of **22.9** μm **26.8** μm and **28.8** μm respectively. The compounds **4 (a)**, **4(c)** and **4 (d)** tested for anti micro (IC_{50}) of the *Fenbendazole* for compounds were found to be **12.7** μm, **10.2** μm and **12.7** μm respectively [45].

4.2 Structure activity relationships (SAR)

The SAR may be attributed from the presence of halo, phenyl and aliphatic linkage present in the benzimidazoles derivatives and deduced from the following points. The minimum inhibitory concentration of the Cl, Br substituted aliphatic amide linkage draw attention for the enhanced binding and increased solubility of the compounds with respect to the target sites. Compounds of benzimidazoles derivatives bearing most active one **4(a)**, **4 (c)** and **4 (d)** as a lead compound to develop novel antimicrobial agent [46]. The appreciable antimicrobial activity of the synthesized Benzimidazole derivatives compared to the standard drugs show that only minor structural changes needed for the derivatives to improve the binding of the molecules. The excellent antimicrobial activity of the synthesized Benzimidazole derivatives compared to the standard drug indicated a fact that for developing novel antimicrobial agent based on synthesized Benzimidazole derivatives. The above results also

Compounds	*E. coli*		*S-Bacillus*		*Staphylococcus A*		*Salmonella Typhi*	
	Zone of Inhibition	**IC_{50}**	**Zone of Inhibition**	**IC_{50}**	**Zone of Inhibition**	**IC_{50}**	**Zone of Inhibition**	**IC_{50}**
	(mm)	**(μm)**	**(mm)**	**(μm)**	**(mm)**	**(μm)**	**(mm)**	**(μm)**
1	1.2	12.9 ± 0.12	3.2	14.8 ± 0.14	4.1	18.9 ± 0.13	7.1	26.9 ± 0.10
2	1.7	15.4 ± 0.18	3.4	18.9 ± 0.15	4.8	25.7 ± 0.11	8.7	22.7± 0.10
3	2.8	22.1 ± 0.17	3.8	25.7 ± 0.12	2.7	22.8 ± 0.02	9.1	24.8 ± 0.11
4a	13.9	29.9 ± 0.12	**12.8**	**51.8** ± 0.19	**2.2**	21.8 ± 0.25	6.9	23.7 ± 0.12
4b	**12.4**	**28.7** ± 0.12	**10.8**	**57.4** ± 0.18	**7.6**	**29.6** ± 0.21	**8.1**	**19.7** ± 0.22
4c	2.8	14.7± 0.15	3.4	15.9± 0.11	4.7	26.9± 0.22	5.4	29.6 ± 0.24
4d	**9.8**	**30.2** ± 0.14	**18.9**	**54.5** ± 0.13	**8.8**	**16.9** ± 0.19	**7.5**	**24.7** ± 0.21
4e	**7.1**	**29.1** ± 0.16	**22.8**	**56.5** ± 0.18	**6.6**	**15.4** ± 0.18	**7.2**	**22.7** ± 0.27
4f	6.8	14.7 ± 0.18	4.1	15.9 ± 0.22	6.9	18.7 ± 0.15	5.5	26.8 ± 0.26
Ciprofloxacin	**18.9**	**48.7** ± 0.01	**17.6**	**51.8** ± 0.01	**14.9**	**42.6** ± 0.01	**16.8**	**43.3**± 0.02

*Potent Antimicrobial Compounds.

Table 1.
Antimicrobial activity data of the synthesized compounds against E. Coli *and* S. Bacillus.

Compounds	*Carbendazim*		*Fenbendazole*	
	Zone of Inhibition (mm)	**IC_{50} (μm)**	**Zone of Inhibition (mm)**	**IC_{50} (μm)**
1	6.1	11.8 ± 0.11	2.2	8.9 ± 0.13
2	9.8	16.1 ± 0.11	2.4	4.7 ± 0.15
3	2.2	10.4 ± 0.10	1.8	18.1 ± 0.14
4a	**7.8**	**22.9** ± 0.25	**1.9**	**12.7** ± 0.18
4b	2.3	14.4 ± 0.23	1.8	13.7± 0.19
4c	**12.2**	**26.8** ± 0.28	**2.2**	**10.2** ± 0.29
4d	**10.2**	**28.8** ± 0.34	**2.7**	**12.7** ± 0.22
4e	2.4	12.5 ± 0.28	2.7	13.8 ± 0.11
4f	2.8	13.8 ± 0.14	5.5	11.8± 0.10
Clotrimazole	12.2	22.8 ± 0.132.2	10.2	26.1 ± 0.13

**Potent antimicrobial compounds.*

Table 2.
Antimicrobial activities of the compounds against Staphylococcus A *and* Salmonella Typhi.

Compounds	*Carbendazim Fenbendazole*				% change in the activity
	Zone of Inhibition (mm)	**IC_{50} (μm)**	**Zone of Inhibition (mm)**	**IC_{50} (μm)**	
2-methyl-1*H*-benzimidazole (**1**)	6.1	11.8 ± 0.11	2.2	8.9 ± 0.13	2.2
1*H*-benzimidazole-2-carboxylic acid (**2**)	9.8	16.1 ± 0.11	2.4	4.7 ± 0.15	3.6
1-(1*H*-benzimidazol-2-yl) methanamine (**3**)	2.2	10.4 ± 0.10	1.8	18.1 ± 0.14	4.8
N-[(1*H*-benzimidazol-2-yl) methyl] thiophene-2-carboxamide (**4a**)	7.8	22.9 ± 0.25	1.9	12.7 ± 0.18*	12.4
N-[(1*H*-benzimidazol-2-yl) methyl]-2-(thiophen-2-yl) acetamide (**4b**)	2.3	14.4 ± 0.23	1.8	13.7± 0.19	1.8
N-[(1*H*-benzimidazol-2-yl) methyl]-3, 6-dichloro-1-benzothiophene -2-carboxamide (**4c**)	12.2	26.8 ± 0.28	2.2	10.2 ± 0.29*	32.4
N-[(1*H*-benzimidazol-2-yl)methyl]-5-(4-fluorophenyl) thiophene-2-carboxamide (**4d**)	10.2	28.8 ± 0.34	2.7	12.7 ± 0.22*	24.7
N-[(1*H*-benzimidazol-2-yl) methyl]-4-bromo-5-chlorothiophene-2-carboxamide (**4e**)	2.4	12.5 ± 0.28	2.7	13.8 ± 0.11	5.9
N-[(1*H*-benzimidazol-2-yl)methyl]-3-chloro-1-benzothiophene-2-carboxamide (**4f**)	2.8	13.8 ± 0.14	5.5	11.8 ± 0.10	3.6
Clotrimazole	12.2	22.8 ± 0.13	10.2	26.1 ± 0.13	1.8

**Potent antifungal agent.*

Table 3.
Antifungal activity data of the synthesized compounds.

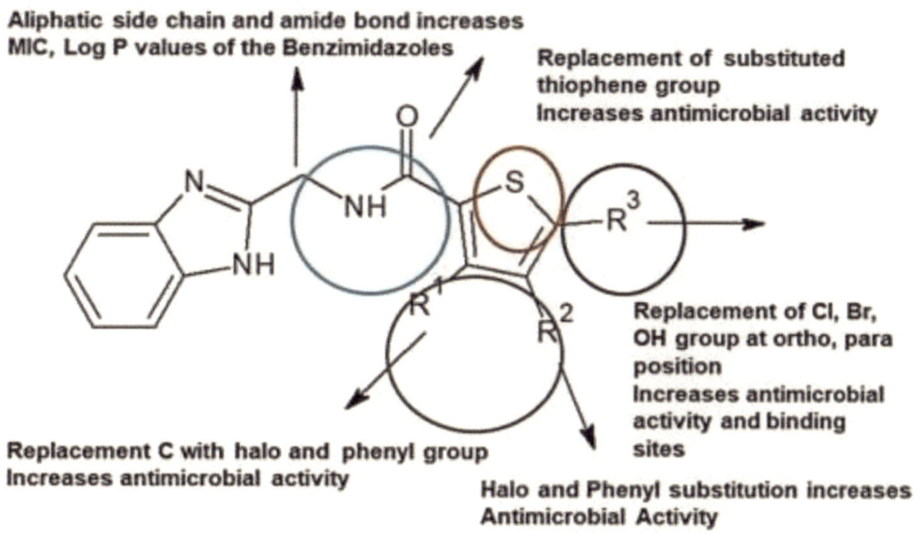

Figure 7.
SAR of the novel derivatives of benzimidazoles containing thiophene moiety.

indicated a fact that different structural requirements are necessary for a compound to show different activities.

SAR study of synthesized compounds showed that the presence of electron-withdrawing moieties phenyl thiophene substituted with Cl, Br groups adjacent to the amide linkage in the benzimidazoles enhanced ant mycobacterial activity [47]. Further SAR of most of the derivatives indicated the attachment of thiophene-2-carboxamide moieties at amide position of the benzimidazoles increased antibacterial activity and the presence of 3, 6-dichloro-1-benzothiophene (**compound 4c**) at 2nd position of Benzimidazole carboximide also important for antimicrobial activities [48, 49]. The presence of electron withdrawing Cl and Br groups at 3rd and 4th position (**compound 4f and 4e**) required for antimicrobial activity. However substitution of 4-fluorophenyl) thiophene at 5th position of benzimidazoles carboxamide moiety increases the solubility of the compound (**compound 4d**) as well as the presence of phenyl ring increases the antibacterial properties (**Figure** 7).

Further, (thiophen-2-yl) acetamide and thiophene-2-carboxamide (**compound 4a and 4b**) attached to the benzimidazoles moiety at 2nd position of the ring system results in basicity of the –NH_2 (amine) linker and thus increases the Clog P values where, C is the concentration of the compounds and P is permeability. On the other hand **compound 4c** contains 3, 6-dichloro-1-benzothiophene-2-carboxamide group attached at 2nd position of the benzimidazoles moiety responsible for moderate antibacterial and antiviral activities. If the 3, 6 dichloro groups present in the opposite direction of the ring system results in enhanced bioavailability of the benzimidazoles derivatives.

5. Conclusion

In the present research, novel series of thiophene amide derivatives containing benzimidazole moiety has been synthesized by the reaction with HATU at room temperature. All the synthesized derivatives are characterized by ^{1}H-NMR, ^{13}C-NMR and LCMS spectroscopic studies. Synthesized compounds are purified by column-chromatography, and tested for antimicrobial strains (antibacterial and antifungal). The cell lines used was *E. coli*, *S. bacillus*, *Staphylococcus Aures* and *Salmonella Typhi* and ciprofloxacin used as standard. The results indicated potential activity towards *S. bacillus* tested compound with IC_{50} of **4 (a)**, **4 (b)**, **4 (d)** and **4 (e)** derivatives against antimicrobial strains with IC_{50} of **51.8** μm, **57.4** μm, **54.5** μm and **56.5** μm respectively. However, compounds **4 (a)**, **4 (c)** and **4 (d)** showed greater inhibitions against *Carbendazim* fungal cell line with IC_{50} of **22.9** μm **26.8** μm and **28.8** μm respectively. Compared with the activity (IC_{50}) of the *Fenbendazole* for compounds **4 (a)**, **4(c)** and **4 (d)** was found to be **12.7** μm, **10.2** μm and **12.7** μm respectively. The newly synthesized derivatives find its potential application in antibacterial and antifungal cells and replacing the existing drug in the market.

Acknowledgements

All the authors are thankful to corresponding author Dr. Pravin Kendrekar, School of Natural Sciences, University of central Lancashire, Preston PR12HE, United Kingdom for carrying out antibacterial and antifungal activity of the synthesized compounds and M. S. Ramaiah University of Applied Science, Karnataka, Bangalore for synthesis of the thiophene substituted benzimidazole derivatives.

Conflict of interest

All the authors declare that they do not have any conflict of interest.

Data availability

All the obtained data has been incorporated in the main manuscript and more data can be obtained from the corresponding author on request.

Author details

Vinayak Adimule[1], Pravin Kendrekar[2]* and Sheetal Batakurki[3]

1 Angadi Institute of Technology and Management (AITM), Belagavi, Karnataka, India

2 Centre for Smart Materials, School of Natural Sciences, University of Central Lancashire, Preston, United Kingdom

3 Department of Chemistry, MS Ramaiah University of Applied Sciences, Bangalore, Karnataka, India

*Address all correspondence to: kkpravin@gmail.com

References

[1] Yogita B, Silakari O. The therapeutic journey of benzimidazoles: A review. Bioorganic & Medicinal Chemistry. 2012;**20**(21):6208-6236. DOI: 10.1016/j.bmc.2012.09.013

[2] Sharma PC, Sinhmar A, Sharma A, Rajak H, Pathak DP. Medicinal significance of benzothiazole scaffold: An insight view. Journal of Enzyme Inhibition and Medicinal Chemistry. 2013;**28**(2):240-266. DOI: 10.3109/14756366.2012.720572

[3] Huigens III, Robert W, et al. The chemical synthesis and antibiotic activity of a diverse library of 2-aminobenzimidazole small molecules against MRSA and multidrug-resistant a. baumannii. Bioorganic & Medicinal Chemistry. 2010;**18**(2):663-674. DOI: 10.1016/j.bmc.2009.12.003

[4] Perin N, Nhili R, Cindrić M, Bertoša B, Vušak D, Martin-Kleiner I, et al. Amino substituted benzimidazo [1,2-a] quinolines: Antiproliferative potency, 3D QSAR study and DNA binding properties. European Journal of Medicinal Chemistry. 2016;**122**:530-545. DOI: 10.1016/j.ejmech.2016.07.007

[5] Starčević K, Marijeta K, Katja E, Ivan S, Magdalena G, Krešimir P, et al. Synthesis, antiviral and antitumor activity of 2-substituted-5-amidino-benzimidazoles. Bioorganic & Medicinal Chemistry. 2007;**15**(13):4419-4426. DOI: 10.1016/j.bmc.2007.04.032

[6] Grogan HM. Fungicide control of mushroom cobweb disease caused by Cladobotryum strains with different benzimidazole resistance profiles. Pest Management Science. 2006;**62**(2): 153-161. DOI: 10.1002/ps.1133

[7] Bhat M, Belagali SL. Guanidinyl benzothiazole derivatives: Synthesis and structure activity relationship studies of a novel series of potential antimicrobial and antioxidants. Research on Chemical Intermediates. 2016;**42**(7): 6195-6208. DOI: 10.1007/s11164-016-2454-6

[8] Shah K, Chhabra S, Shrivastava SK, Mishra P. Benzimidazole: A promising pharmacophore. Medicinal Chemistry Research. 2013;**22**(11):5077-5104. DOI: 10.1007/s00044-013-0476-9

[9] Demeunynck M, Bailly C, Wilson WD, editors. DNA and RNA Binders, from Small Molecules to Drugs. Vol. 2. Weinheim, Germany: Wiley-VCH; 2006. ISBN: 978-3-527-60566-8

[10] Racané L, Tralic-Kulenovic V, Pavelić SK, Ratkaj I, Peixoto P, Nhili R, et al. Novel diamidino-substituted derivatives of phenyl benzothiazolyl and dibenzothiazolyl furans and thiophenes: Synthesis, antiproliferative and DNA binding properties. Journal of Medicinal Chemistry. 2010;**53**(6):2418-2432. DOI: 10.1021/jm901441b

[11] Hranjec M, Sović I, Ratkaj I, Pavlović G, Ilić N, Valjalo L, et al. Anti-proliferative potency of novel benzofuran-2-carboxamides on tumour cell lines: Cell death mechanisms and determination of crystal structure. European Journal of Medicinal Chemistry. 2013;**59**:111-119. DOI: 10.1016/j.ejmech.2012.11.009

[12] Adimule V, Medapa S, Kumar LS, Rao PK. Novel substituted phenoxy derivatives of 2-chloro-n-{5-[2-(4-methoxy-phenyl)-pyridin-3-yl]-[1, 3, 4] thiadiazol-2-yl}-acetamides: Synthesis, characterization and in-vitro anticancer properties. Journal of Pharmaceutical and Chemical Biological Sciences. 2014;**2**(2):130-137

[13] Vinayak A, Sudha M, Lalita KS. Design, synthesis and characterization of novel amine derivatives of 5-[5-(Chloromethyl)-1, 3, 4-Oxadiazol-2-yl]-2-(4-Fluorophenyl)-pyridine as a new class of anticancer agents. Dhaka University Journal of Pharmaceutical Sciences. 2017;**16**(1):11-19. DOI: 10.3329/dujps.v16i1.33377

[14] Vinayak A, Sudha M, Jagadeesha A, Lalita K. Design, synthesis, characterization and cancer cell growth-inhibitory properties of novel derivatives of 2-(4-Fluoro-phenyl)-5-(5-aryl Substituted-1, 3, 4-Oxadiazol-2-yl) pyridine. Journal of Pharmaceutical Research International. 2015;**7**(1):34-43. DOI: 10.9734/BJPR/2015/15486

[15] Adimule VM et al. Design, synthesis and cytotoxic evaluation of novel 2-(4-N, N-dimethyl) pyridine containing 1, 3, 4-oxadiazole moiety. Asian Journal of Biomedical and Pharmaceutical Sciences. 2014;**1**:4-37

[16] Basavaraja KM, Somasekhar B, Shivakumar B. Synthesis of 2-[(1-phenyl)(aryl) azo] methyleneimino-6-chloro/fluoro benzothiazoles and their antibacterial activity. International Journal of PharmTech Research. 2010; **2**(2):1139-1143

[17] Perin N, Martin-Kleiner I, Nhili R, Laine W, David-Cordonnier M-H, Vugrek O, et al. Biological activity and DNA binding studies of 2-substituted benzimidazo [1, 2-a] quinolines bearing different amino side chains. Medicinal Chemistry Communication. 2013;**4**(12): 1537-1550. DOI: 10.1039/C3MD00193H

[18] Mirela S, Poznic M, Gehrig P, Scott M, Schlapbach R, Hranjec M, et al. Differential antiproliferative mechanisms of novel derivative of benzimidazo [1, 2-α] quinoline in colon cancer cells depending on their p53 status. Molecular Cancer Therapeutics. 2008;**7**(7):2121-2132. DOI: 10.1158/1535-7163.MCT-07-2261

[19] Liu H, Sihui Long KP, Rakesh G-FZ. Structure-activity relationships (SAR) of triazine derivatives: Promising antimicrobial agents. European Journal of Medicinal Chemistry. 2020;**185**: 111804. DOI: 10.1016/j.ejmech.2019.111804

[20] Nimesh H, Sur S, Sinha D, Yadav P, Anand P, Bajaj P, et al. Synthesis and biological evaluation of novel bisbenzimidazoles as Escherichia coli topoisomerase IA inhibitors and potential antibacterial agents. Journal of Medicinal Chemistry. 2014;**57**(12): 5238-5257. DOI: 10.1021/jm5003028

[21] Agh-Atabay NM, Dulger B, Gucin F. Structural characterization and antimicrobial activity of 1, 3-bis (2-benzimidazyl)-2-thiapropane ligand and its Pd (II) and Zn (II) halide complexes. European Journal of Medicinal Chemistry. 2015;**40**(11): 1096-1102. DOI: 10.1016/j.ejmech.2005.05.006

[22] Soderlind K-J, Gorodetsky B, Singh AK, Bachur NR, Miller GG, William Lown J. Bis-benzimidazole anticancer agents: Targeting human tumour helicases. Anti-Cancer Drug Design. 1999;**14**(1):19-36

[23] Al-Mudaris ZA, Majid ASA, Ji D, Al-Mudarris BA, Chen S-H, Liang P-H, et al. Conjugation of benzylvanillin and benzimidazole structure improves DNA binding with enhanced antileukemic properties. PLoS One. 2018;**8**(11): e80983. DOI: 10.1371/journal.pone.0080983

[24] Nasser MY, Alatibi F, El-Sayed NNE, Al-Showiman S, Kheder NA, Wadood A, et al. Antimicrobial activity of some

novel armed thiophene derivatives and petra/osiris/molinspiration (POM) analyses. Molecules. 2016;**21**(2):222. DOI: 10.3390/molecules21020222

[25] Mishra R, Sharma PK, Verma P, Mishra I. Antimicrobial potential of thiophene derivatives of synthetic origin: A review. European Chemical Bulletin. 2016;**5**:399-407

[26] Fokialakis N, Cantrell CL, Duke SO, Skaltsounis AL, Wedge DE. Antifungal activity of thiophenes from Echinops ritro. Journal of Agricultural and Food Chemistry. 2006;**54**(5):1651-1655. DOI: 10.1021/jf052702j

[27] Pinto E, Queiroz M-JRP, Vale-Silva LA, Oliveira JF, Begouin A, Begouin J-M, et al. Antifungal activity of synthetic di (hetero) arylamines based on the benzo [b] thiophene moiety. Bioorganic & Medicinal Chemistry. 2008;**16**(17): 8172-8177. DOI: 10.1016/j. bmc.2008.07.042

[28] El-Gaby MSA, Gaber AM, Atalla AA, Al-Wahab KAA. Novel synthesis and antifungal activity of pyrrole and pyrrolo [2,3-d] pyrimidine derivatives containing sulfonamido moieties. Farmaco. 2002;**57**(8):613-617. DOI: 10.1016/S0014-827X(01)01178-8

[29] Refat HM, Fadda AA. Synthesis and antimicrobial activity of some novel hydrazide, benzochromenone, dihydropyridine, pyrrole, thiazole and thiophene derivatives. European Journal of Medicinal Chemistry. 2013;**70**:419-426. DOI: 10.1016/j.ejmech.2013.09.003

[30] Nakib TAI, Meegan MJ, Looney AM, Burke ML. Synthesis and antifungal activity of some 2-aryl-3-hydroxymethylbenzo [b]thiophenes. European Journal of Medicinal Chemistry. 1992;**27**(9):971-976. DOI: 10.1016/0223-5234(92)90031-U

[31] Adimule V, Sudha M, Lalita KS, Kumar RP. Synthesis, characterization and cytotoxic evaluation of novel derivatives of 1-[2-(aryl substituted)-5-(4′-Fluoro-3-methyl biphenyl-4-yl)-[1, 3, 4] oxadiazole-3-yl]-ethanone. Archives of Applied Science Research. 2015;**7**(5):4-8

[32] Vinayak A, Sudha M, Sanjeev KL, Kumar RP. Design, synthesis, characterization and anticancer properties of novel 2-chloro-N-(aryl substituted) acetamide derivatives of 5-[2-(4-methoxyphenyl) pyridin-3-yl]-1, 3, 4-oxadiazole-2-thiol. International Journal of Drug Development and Research. 2014;**6**(4):188-195

[33] Adimule V, Medapa S, Adarsha HJ, Kumar SL, Rao PK. Synthesis, characterization and in vitro anticancer properties of 1-{5-Aryl-2-[5-(4-Fluoro-phenyl)-Thiophen-2-yl]-[1, 3, 4] Oxadiazol-3-yl}-Ethanone. International Journal of Pharma Research & Review. 2014;**3**(12):20-25

[34] Vinayak A, Sudha M, Kumar RP, Kumar Lalita S. Synthesis, characterization and anticancer activity of Schiff base derivatives of 5-(2-phenoxypyridin-3-yl)-1, 3, 4-thiadiazol-2-amine. International Research Journal of Pharmacy. 2013;**4**(12):62-66

[35] Adimule V, Sudha M, Lalita KS, Kumar RP. Synthesis, characterization and cytotoxic evaluation of novel derivatives of 1-[2-(aryl substituted)-5-(4'-Fluoro-3-methyl biphenyl-4-yl)-[1, 3, 4] oxadiazole-3-yl]-ethanone. Archives of Applied Science Research. 2015:**7**(5):4-8

[36] Adimule V, Sudha M, Lalita KS, Kumar R. Design, synthesis, characterization and cytotoxic evaluation of novel 2-chloro-n-(aryl substituted) acetamide derivatives of 5-

[2-phenyl pyridin-3-yl]-1, 3, 4-oxadiazole-2-thiol. Research Journal of Pharmaceutical, Biological and Chemical Sciences. 2014;**2**(2):130-137

[37] Tokura S, Ueno K, Miyazaki S, Nishi N. Molecular weight dependent antimicrobial activity by chitosan. In: New Macromolecular Architecture and Functions. Berlin, Heidelberg: Springer; 1996. DOI: 10.1007/978-3-642-80289-8_21

[38] Rhim J-W, Hong S-I, Park H-M, Ng PKW. Preparation and characterization of chitosan-based nanocomposite films with antimicrobial activity. Journal of Agricultural and Food Chemistry. 2006;**54**(16):5814-5822. DOI: 10.1021/jf060658h

[39] Pantić DN, Aranđelović S, Radulović S, Roller A, Vladimir BA, Grgurić-Šipka S. Synthesis, characterization and cytotoxic activity of organoruthenium (II)-halido complexes with 1H-benzimidazole-2-carboxylic acid. Journal of Organometallic Chemistry. 2016;**819**:61-68. DOI: 10.1016/j.jorganchem

[40] Machura B, Wolff M, Benoist E, Schachner JA, Mösch-Zanetti NC, Takao K, et al. Oxorhenium (V) complexes with 1H-benzimidazole-2-carboxylic acid-synthesis, structural characterization and catalytic application in epoxidation reactions. Polyhedron. 2014;**69**:205-218. DOI: 10.1016/j.poly.2013.12.004

[41] Rees LP, Minney SF, Plummer NT, et al. A quantitative assessment of the antimicrobial activity of garlic (*Allium sativum*). World Journal of Microbiology and Biotechnology. 1993;**9**:303-307. DOI: /10.1007/BF00383068

[42] Balouiri M, Sadiki M, Ibnsouda SK. Methods for in vitro evaluating antimicrobial activity: A review. Journal of Pharmaceutical Analysis. 2016;**6**(2): 71-79. DOI: 10.1016/j.jpha.2015.11.005

[43] Greenwalt CJ, Ledford RA, Steinkraus KH. Determination and characterization of the antimicrobial activity of the fermented tea Kombucha. LWT-Food Science and Technology. 1998;**31**(3):291-296. DOI: 10.1006/fstl.1997.0354

[44] Remmal A, Bouchikhi T, Rhayour K, Ettayebi M, Tantaoui-Elaraki A. Improved method for the determination of antimicrobial activity of essential oils in agar medium. Journal of Essential Oil Research. 1993;**5**(2):179-184. DOI: 10.1080/10412905.1993.9698197

[45] Kowalska-Krochmal B, Dudek-Wicher R. The minimum inhibitory concentration of antibiotics: Methods, interpretation, clinical relevance. Pathogens (Basel, Switzerland). 2021; **10**(2):165. DOI: 10.3390/pathogens10020165

[46] de Oliveira HC et al. Fenbendazole controls in vitro growth, virulence potential, and animal infection in the Cryptococcus model. Antimicrobial Agents and Chemotherapy. 2020;**64**(6): e00286-e00220. DOI: 10.1128/AAC.00286-20

[47] Tafere Mulaw Belete. Novel targets to develop new antibacterial agents and novel alternatives to antibacterial agents. Human Microbiome Journal. 2019;**11**: 2452-2317. DOI: 10.1016/j.humic.2019.01.001

[48] Umesh K, Rakesh N, Surendra KN, Sachin KS, Vivek G. Benzimidazole: Structure activity relationship and mechanism of action as antimicrobial gent. Research Journal of Pharmacy and Technology. 2017;**10**(7):2400-2414. DOI: 10.5958/0974-360X.2017.00425.5

[49] Brishty SR, Hossain MJ, Khandaker MU, Faruque MRI, Osman H, Rahman SMA. A comprehensive account on recent progress in pharmacological activities of benzimidazole derivatives. Frontiers in Pharmacology. 2021;**12**:762807. DOI: 10.3389/fphar.2021.762807

Section 3

Spectral and Theoretical Study of Benzimidazole

Chapter 6

Spectral and Theoretical Studies of Benzimidazole and 2-Phenyl Substituted Benzimidazoles

A. Antony Muthu Prabhu

Abstract

This chapter discusses about the spectral and theoretical aspects of selected benzimidazole and 2-phenyl substituted benzimidazole molecules. The synthesis of these benzimidazoles was reported in many methods by the reaction between o-phenylenediamine with formic acid, aromatic aldehydes and N-benzylbezene-1,2-diamine in presence of oxidant tert-butyl hydroperoxide (TBHP). The spectral analysis of these molecules mainly such as UV-visible, fluorescence in solvents will be included in this chapter and discussed about the absorption, fluorescence maximum, conjugation, transition. Further the optimized structure of these molecules will be given using Gaussian 09 W (DFT 6-31G method). And also will be discussed about structural parameters, highest occupied molecular orbital (HOMO) – lowest unoccupied molecular orbital (LUMO) energy energy values, natural bond orbital (NBO), molecular electrostatic potential map (ESP). Many benzimidazole molecules having tautomers in the structure will be explained with the help of theoretical parameters to describe the structural properties.

Keywords: benzimidazole, spectral properties, computational study, NBO, MSP

1. Introduction

This chapter will be discussed about the spectral and theoretical studies of benzimidazole and 2-phenyl substituted benzimidazoles. A series of benzimidazole molecules are very important heterocyclic compounds in organic chemistry with two nitrogen atoms in five membered ring fused with aromatic moiety having different nature in spectral and biological properties. Benzimidazoles which contain a hydrogen atom attached to nitrogen in the 1-position readily tautomerize. In this way, many benzimidazole molecules are synthesized from the basic moiety to involving in large applications especially in medicinal fields. This type of benzimidazole derivatives possess the many pharmaceutical properties such as antiviral [1], antitumor [2], antihistaminic [3], antimicrobial [4], antihelminthic [5, 6], anticancer [7], antifungal [8], antimicrobial [4], antibacterial [9], analgesic [10], anti-convulsant [11] and anti-ulcer [12] activity. Some of benzimidazole molecules are used as corrosion inhibitors for metals and alloy surfaces in industrial field [13–15].

Particularly, Albendazole, Mebendazole and Thiabendazole having benzimidazole moiety are widely used as anthelmintic drugs [16].

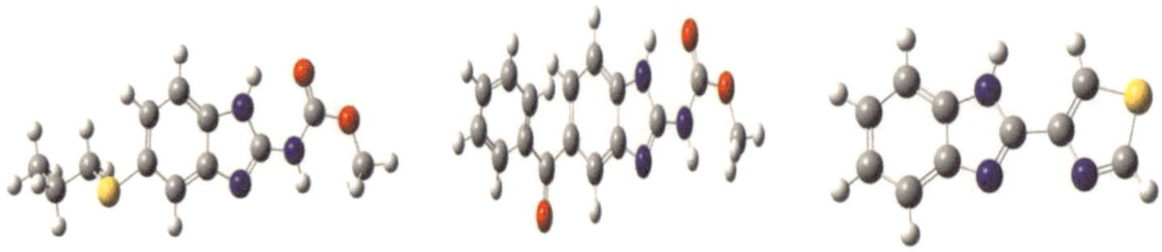

Albendazole Mebendazole Thiabendazole

Some fluoroquinolones substituted benzimidazole derivatives have been reported by microwave assisted method. The synthesized compounds are reported to be the derivatives of Ciprofloxacin & Norfloxacin [17].

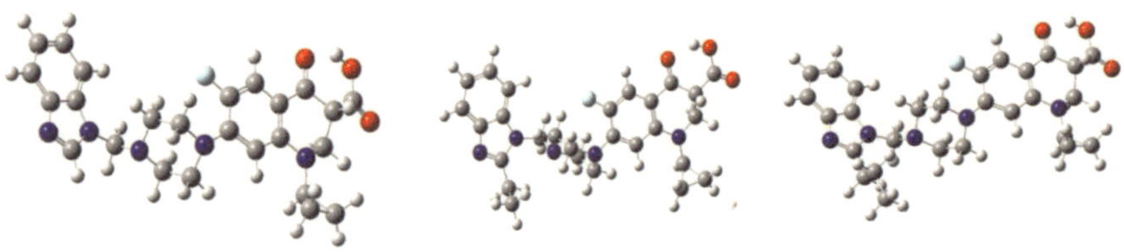

Ciprofloxacin derivatives

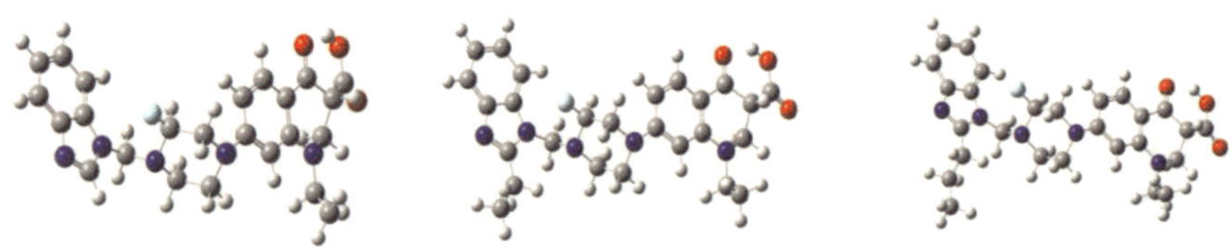

Norfloxacin derivatives

The structural studies of synthesized benzimidazole derivatives are characterized using the spectral techniques such as single crystal XRD, UV-visible, Infrared, ^{1}H NMR, ^{13}C NMR etc. [18–21]. These main techniques are usually referred for characterizing many organic synthesized molecules to elucidate the presence of functional groups, conjugation and structural parameters. Particularly the absorption and fluorescence spectral properties of these benzimidazole derivatives have been changed with respect to the change in substitution in aromatic ring at o- and p-position [22–28].

Another important application of benzimidazoles is involved to exhibit the excited state intra-molecular proton transfer reaction (ESIPT) [22, 25, 29–31]. Particularly, the presence of hydroxyl group in benzimidazole at 2-position in benzene ring is exhibited this process through intramolecular hydrogen bonding between the acidic protons (-OH, -NH_2) and basic centers (=N-, -C=O) in same molecule. ESIPT process for benzimidazole molecules is observed through dual fluorescence in aqueous solvent, one a normal stoke shifted fluorescence band and second large stoke shifted fluorescence band. Absorption and emission spectral study of these molecules were reported in different solvents with changing polarity. Moreover, the 2(2′-hydroxyphenyl)benzimidazole molecule are studied the enhancement of ESIPT process in aqueous β-cyclodextrin through the formation of host-guest inclusion complex [32].

The density functional theory (DFT) studies give information regarding the structural parameters, the functional groups, orbital interactions and vibrational frequencies [33]. The DFT calculations with the hybrid exchange-correlation functional B3LYP (Becke's three parameter (B3) exchange in conjunction with the Lee–Yang–Parr's (LYP) correlation functional) which are especially important in systems containing extensive electron conjugation and/or electron lone pairs [34–36]. The HOMO–LUMO energy, MSP map and the Mullikan population analysis will be calculated for the studied molecules. The natural bond orbital (NBO) analysis will explains the most important orbital interactions in order to clarify general structural features. The excited state potential energy surface, excited state intramolecular proton transfer of 2-(2′-Hydroxyphenyl)benzimidazole was investigated by TD-DFT method in gas phase and in solvent [37–39]. Further the theoretical calculations of Methyl-6-Nitro-1H-Benzimidazole and 1-methyl-2-phenyl benzimidazole was reported [40, 41]. Many benzimidazole molecules were reported theoretically structural properties and vibrational spectra, HOMO-LUMO, NBO analysis by ab initio HF and DFT methods [42–45]. Theoretical calculations of many benzimidazole molecules in gas phase were analyzed for the structural investigation with the help of experimental results [20, 44, 46–49].

2. Benzimidazole and 2-phenyl benzimidazoles

The benzimidazole molecule without any substitution was synthesized by the simple reaction between o-phenylenediamine with formic acid in the presence of alkali like sodium hydroxide, potassium hydroxide etc., [50]. The synthesis procedure of benzimidazole is given briefly. 27 g of o-phenylenediamine and 17.5 g of 90% formic acid are taken a 250 ml round bottom flask, the reaction mixture is heated at 100°C on a water bath for 2 hours. After cooling this reaction mixture, the 10% sodium hydroxide solution is added slowly to the solution with constant rotation then the reaction mixture becomes alkaline. Crude benzimidazole is filterd off at the pumb and washed with 25 ml of cold water and the crude product is dissolved in 400 ml of boiling water. Then 2 g of decolourising carbon is added to the solution and heat for 20 minutes. Finally the benzimidazole is formed after filteration at the pumb. New series of benzimidazole and its derivatives were synthesized and characterized using spectral analysis and applied for biological properties [51, 52]. For example, the series of substituted benzimidazole were reported particularly in the 2-position and 1-position replacing the hydrogen atom by both small and large size molecules (**Figure 1**).

The substituted 2-phenyl benzimidazoles were synthesized from the condensation reaction between *o*-Phenylenediamine and substituted aromatic aldehydes in chloroform and in the presence of ammonium chloride as a catalyst [53]. This reaction carried out at room temperature using many ammonium salts like ammonium bromide (NH_4Br), ammonium chloride (NH_4Cl), ammonium fluoride (NH_4F), ammonium sulphate ($(NH_4)_2SO_4$) and ammonium carbonate ($(NH_4)_2CO_3$). Aromatic aldehydes such as benzaldehyde, m-methyl benzaldehyde, p-methyl benzaldehyde,

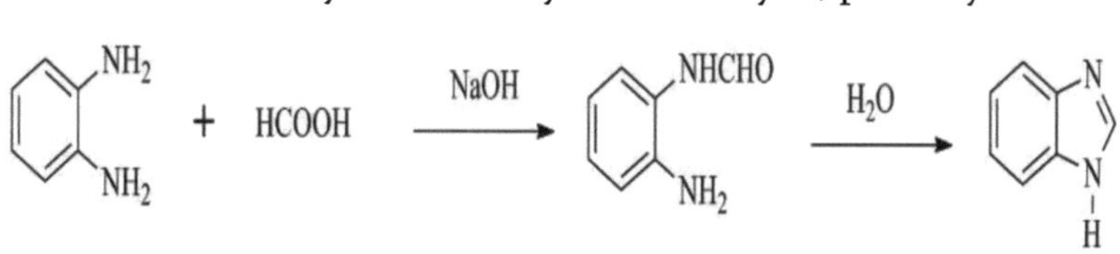

Figure 1.
The reaction shows between o-phenylene diamine with formic acid in presence of sodium hydroxide.

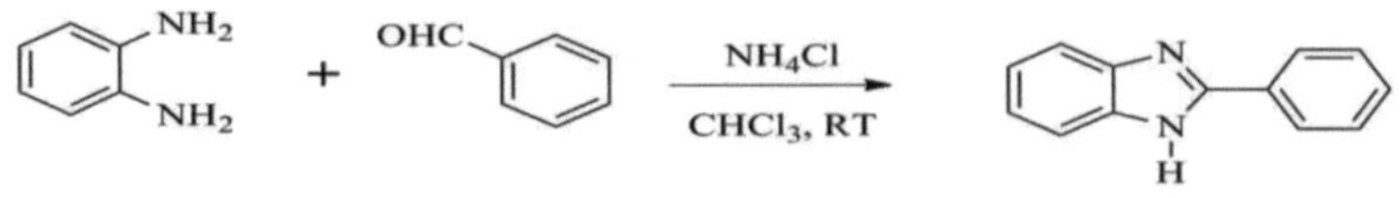

Figure 2.
The reaction shows between o-phenylene diamine with aromatic aldehyde in chloroform and in presence of ammonium chloride.

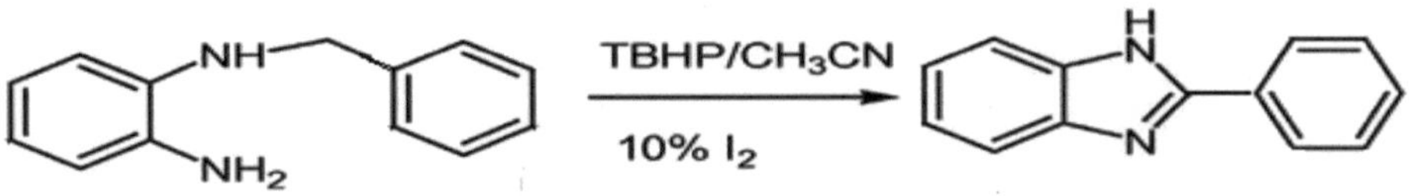

Figure 3.
Tandem oxidative cyclization of different N-substituted benzene-1,2-diamines.

m-methoxy benzaldehyde, p-methoxy benzaldehyde, o-hydroxy benzaldehyde, p-hydroxy benzaldehyde, o-amino benzaldehyde, p-amino benzaldehyde are used for the preparation of 2-phenyl substituted benzimidazole molecules. 1,2-phenylenediamine is added to the stirred solution of aromatic aldehydes and NH_4Cl in $CHCl_3$. This reaction mixture is stirred for 4 hours at room temperature. After completion of the reaction, the phenyl substituted benzimidazoles are formed using Thin layer chromatography, column chromatography (**Figure 2**).

Thus 2-phenyl substituted benzimidazole derivative has been reported by another method and the reaction scheme shows in **Figure 3**. The oxidative dehydrative coupling reaction of N-benzylbenzene-1,2-diamine in the presence of oxidant tert-butyl hydroperoxide (TBHP) in solvent acetonitrile to give substituted 2-phenyl benzimidazoles [54]. N-benzylbenzene1,2-diamine 1.96 g, I_2 0.25 g, TBHP 1.8 g, is added in a 25 ml round bottomed flask in acetonitirle solvent and stirred at room temperature. The product is purified by column chromatography and finally the phenyl substituted benzimidazole is formed at the end of reaction.

3. Absorption and emission spectral study

Thus, the molecular diagrams of studied benzimidazole derivatives are shown in **Figure 4**. In this section, the absorption and emission spectral study is discussed for the selected benzimidazoles in solvents. The absorption and emission maximum was observed at 273 nm, 279 nm and at 291 nm for benzimidazole [55, 56] and at 295 and 350 nm for 2-phenyl benzimidazole [57]. These maximum are similar to 2-(m-methylphenyl)benzimidazole, 2-(p-methylphenyl)benzimidazole, 2-(m-methoxyphenyl)benzimidazole, 2-(p-methoxyphenyl)benzimidazole. But the 2-(o-hydroxyphenyl)benzimidazole molecule is observed the specific property in the ground and excited states to describe the keto-enol tautomeric equilibrium through the excited state intramolecular proton transfer [58–65], which property already given in introduction part and theoretical study also done for keto-enol tautomer in solvent effect. The absorption maximum of 2-(o-hydroxyphenyl)benzimidazole was observed at 335, 318 and 295 nm and fluorescence maximum at 355 and 465 nm in DMSO [32]. The absorption and emission maximum was observed at 320, 285 and 240 nm and fluorescence maximum at 417 nm in water for 2-(o-aminophenyl)benzimidazole [66] and at 313, 255, and 207 nm and at 382 nm in water for 2-(p-aminophenyl) benzimidazole [67].

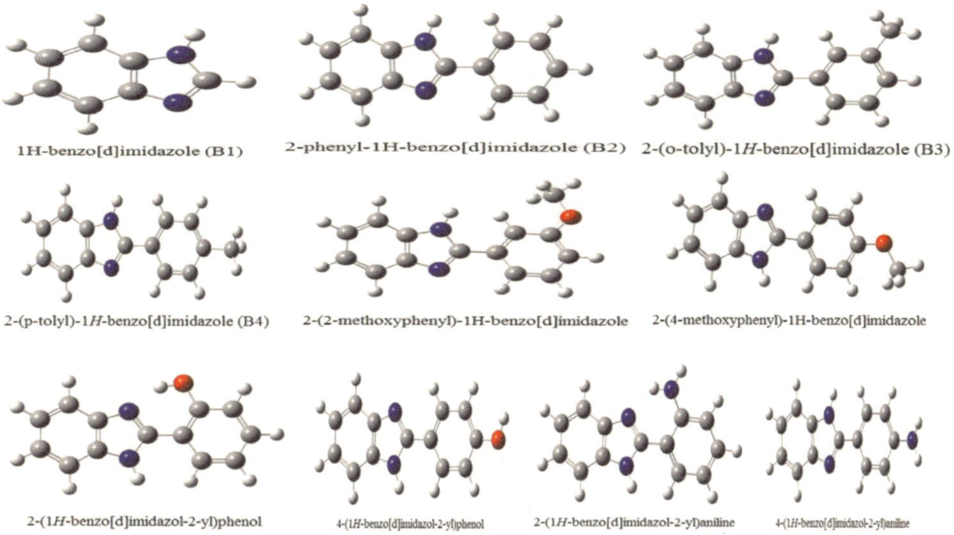

Figure 4.
It shows the molecular diagram of selected molecules by DFT B3LYP 6-31G method.

4. Theoretical study

4.1 HOMO-LUMO energy parameters

Initial computational study of selected benzimidazoles has been investigated from the HOMO, LUMO energy diagram in gas phase using DFT B3LYP 6-31G method. The chemical stability of selected benzimidazoles is demonstrated with the help of explaining this energy diagram [68]. Generally, the HOMO energy picture represents the ability to donate an electron and LUMO energy picture represent the ability to obtain an electron. For selected benzimidazoles the electron density is completely localized on the whole ring for both orbitals which shows in **Table 1**. The higher HOMO-LUMO energy gap of benzimidazole (−5.56 eV) and lower HOMO-LUMO energy gap of 2-(o-aminophenyl)benzimidazole (−4.38 eV) is observed theoretically to describing the stability and reactivity with addition of substitution in the phenyl ring. If the substitution in the benzimidazole is clearly changed the HOMO-LUMO energy gap particularly the amino group at o-position in the phenyl ring with lower value.

The theoretical physical parameters of selected benzimidazoles are determined by electronic chemical potential (μ), electronegativity (χ), absolute hardness (η), softness (S) and electrophilicity (ω) values from the HOMO as ionization energy (IE) and LUMO as electron affinity (EA) using the following equations, respectively. These parameters can be calculated using the following Eqs. (1)–(5).

$$\mu = (E_{HOMO} + E_{LUMO})/2 \tag{1}$$

$$\chi = -\mu \tag{2}$$

$$\eta = (E_{HOMO} - E_{LUMO})/2 \tag{3}$$

$$S = 1/\eta \tag{4}$$

$$\omega = \mu^2/2\eta \tag{5}$$

S. No.	Compounds name	HOMO diagram	LUMO diagram	MSP diagram
1	Benzimidazole			
2	2-phenyl benzimidazole			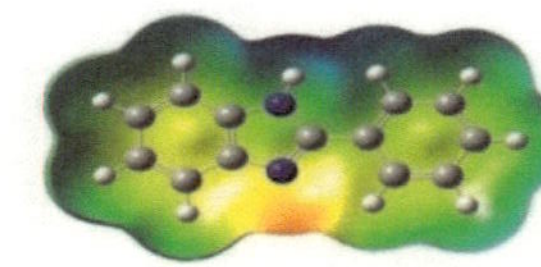
3	2-(m-methylphenyl) benzimidazole			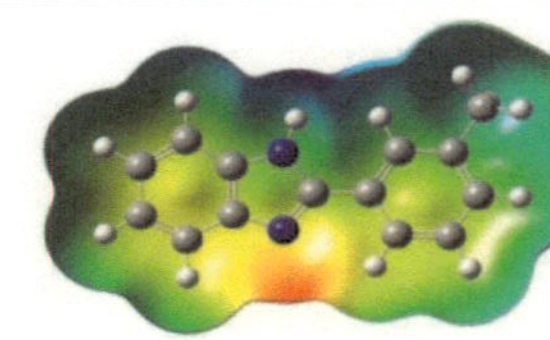
4	2-(m-methoxyphenyl) benzimidazole			
5	2-(p-methylphenyl) benzimidazole			

S. No.	Compounds name	HOMO diagram	LUMO diagram	MSP diagram
6	2-(p-methoxyphenyl) benzimidazole			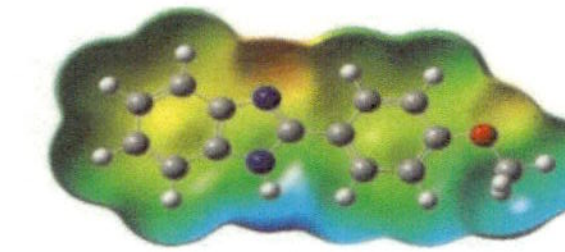
7	2-(o-hydroxyphenyl) benzimidazole			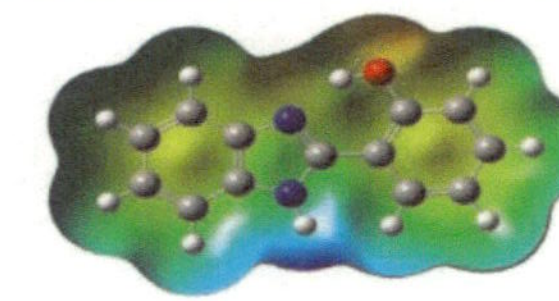
8	2-(p-hydroxyphenyl) benzimidazole			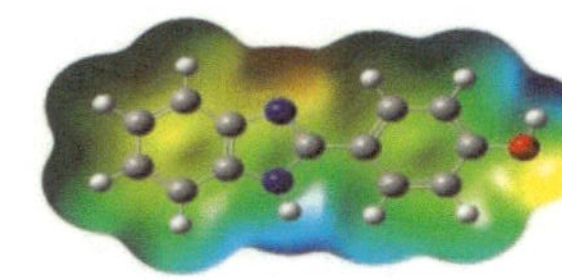
9	2-(o-aminophenyl) benzimidazole			
10	2-(p-aminophenyl) benzimidazole			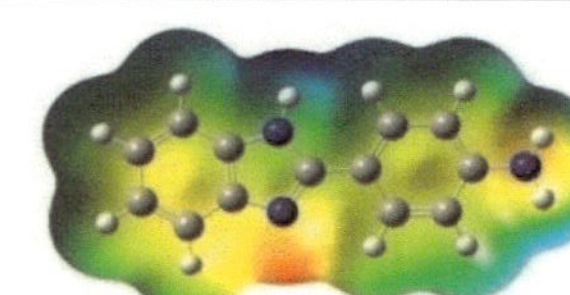

Table 1.
HOMO, LUMO energy diagram and molecular electrostatic potential (MSP) diagram obtained by DFT B3LYP 6-31G method.

DOI: http://dx.doi.org/10.5772/intechopen.101966

Theoretically calculated absolute hardness and softness are observed in the range from 1.11 to 2.78 eV. The values of μ, χ, η, S, ω for the amino group at o- and p-position leads to lesser values due to the electron donating nature of amino group. Absolute hardness and softness are important properties to measure the molecular stability and reactivity. S has been known as an indicator of the overall stability of a chemical system. A hard molecule has a large energy gap and a soft molecule has a narrow energy gap. Soft molecules are more reactive than hard molecules because they could easily offer electrons to an acceptor. For the simplest transfer of electrons, adsorption could occur at the part of the molecule where softness has the highest magnitude and hardness has the lowest [69]. The absolute hardness and softness of Albendazole molecule (benzimidazole moiety) was reported in the values of 2.56 and 0.39 eV in the gas phase [70]. Further the parameter of electrophilicity index (ω) of a organic molecule gives the information about the binding ability with biomolecules [71–73].

The calculated dipole moment values of the methoxy substituent at m- and p- are higher than those of other derivatives. But the hydroxyl group at o-position is higher than that of p-position and in amino group at p-position is higher value than that of o-position. The compound that has the highest dipole moment value is 2-(-p-methoxyphenyl)benzimidazole with the value of 4.91 D. 2-(p-hydroxyphenyl) benzimidazole has the lowest dipole moment among the studied compounds (1.88 D).

Theoretically calculated energy for selected benzimidazole molecules are investigated with respect to substitution at m- and p-position of methyl, methoxy and also o- and p-position for hydroxyl and amino groups. More negative energy values are observed for 2-(m-methoxyphenyl)benzimidazole and 2-(p-methoxyphenyl)benzimidazole. Further the comparisons of substitutions at m-, p-positions and o-, p-position are not significantly changed in energy values in gas phase. Also thermodynamic parameters such as enthalpy, free energy and entropy are calculated theoretically at room temperature in the gas phase. All values of substituted benzimidazoles are higher than the parent benzimidazole (**Table 2**).

4.2 Molecular electrostatic potential

Another investigating theoretical study is molecular electrostatic potential map (MSP) and displayed in **Table 1**. The electrophiles tend to negative ESP and the nucleophiles tend to region of positive ESP. The molecular electrostatic potential was calculated with DFT B3LYP 6-31G level of theory. The negative regions (red) are mainly contained on the nitrogen atom (=N-) and oxygen atom from methoxy, hydroxyl groups while the positive regions (blue) for the proton from –N-H, -OCH_3, -OH and -NH_2 group.

4.3 Natural bond orbital (NBO) analysis

Further to study of the intramolecular interactions of selected benzimidazoles, the theoretical NBO is used to calculate the stabilization energy for -C-H, -N-H and lone pair electrons in heteroatoms using second-order perturbation theory [74]. Some calculations about NBO analysis for investigating the intra and intermolecular interactions of isolated organic molecules and inclusion complexes between organic molecules and cyclodextrins [75, 76] were reported theoretically. NBO 3.1 program is applied to perform the natural bond orbital (NBO) analysis [77, 78] in the Gaussian 09 W package at the DFT/B3LYP level. In this work, DFT B3LYP 6-31G method is applied to analysis of intramolecular interactions for selected benzimidazoles. The

Parameters	Benz imidazole	2-phenyl benz imidazole	2-(m-methyl phenyl) benz imidazole	2-(m-methoxy phenyl) benz imidazole	2-(p-methyl phenyl) benz imidazole	2-(p-methoxy phenyl) benz imidazole	2-(o-hydroxy phenyl) benz imidazole	2-(p-hydroxy phenyl) benz imidazole	2-(o-amino phenyl) benz imidazole	2-(p-amino phenyl) benz imidazole
EHOMO (eV)	−6.07	−5.84	−5.79	−5.74	−5.73	−5.59	−5.72	−5.62	−5.37	−5.30
ELUMO (eV)	−0.50	−1.17	−1.12	−1.15	−0.92	−0.98	−1.25	−1.01	−0.99	−0.81
EHOMO-ELUMO (eV)	−5.56	−4.66	−4.66	−4.59	−4.80	−4.60	−4.47	−4.61	−4.38	−4.49
μ (eV)	−3.28	−3.50	−3.45	−3.44	−3.32	−3.28	−3.48	−3.31	−3.18	−3.05
χ (eV)	3.28	3.50	3.45	3.44	3.32	3.28	3.48	3.31	3.18	3.05
η (eV)	2.78	2.33	2.33	2.29	2.40	2.30	2.23	2.30	2.19	2.24
S (eV)	1.39	1.16	1.16	1.14	1.20	1.15	1.11	1.15	1.09	1.12
ω (eV)	1.93	2.62	2.54	2.57	2.29	2.33	2.70	2.37	2.30	2.07
μ (Debye)	3.61	3.21	3.59	4.05	3.43	4.91	4.07	1.88	1.92	4.33
E (kcal mol^{-1})	−238299.56	−383263.72	−407929.73	−455101.17	−407929.73	−455102.22	−428199.43	−430445.47	−417989.88	−417987.46
G (kcal mol^{-1})	56.89	104.71	119.60	123.60	120.89	123.44	107.09	106.93	114.97	113.24
H (kcal mol^{-1})	80.00	133.89	152.48	155.57	152.41	155.55	137.32	136.53	145.56	144.43
S (kcal mol^{-1} K^{-1})	77.53	97.85	110.26	107.22	105.71	107.71	101.39	99.26	102.60	104.61

Table 2.
The values of energy of HOMO, LUMO, HOMO-LUMO energy gap, structural parameters, dipole, energy, thermodynamic parameters of selected benzimidazoles obtained by DFT B3LYP 6-31G method.

stabilization energy $E^{(2)}$ for the donor and acceptor orbital delocalization is involved through the occupied Lewis-type (bond or lone pair) NBOs and formally unoccupied (antibond or Rydberg) non-Lewis NBOs within the molecules [79, 80]. The electron density of all atoms is noted in the selected benzimidazole molecules.

The second-order Fock matrix is carry out to evaluate the donor–acceptor interactions within the molecules to specify in conjugative π bonds and lone pair electrons through the NBO analysis [81]. The Eq. (6) is given below to estimate the stabilization energy $E^{(2)}$ for the donor and acceptor orbital delocalization within the molecule.

$$E^{(2)} = \Delta E_{ij} = q_i \frac{F(i,j)^2}{\varepsilon_j - \varepsilon_i} \tag{6}$$

where q_i is the donor orbital occupancy, ε_i and ε_j are diagonal elements (orbital energies), and $F(i, j)$ is the off-diagonal NBO Fock matrix element [82–84]. This analysis reveals that the conjugative interaction, hyper-conjugative interaction, intra and intermolecular hydrogen bond in the same molecules and combine with other molecules are well described for the donor – acceptor orbitals.

Thus the interaction between π(N11-C12) and π*(C4-C7) are reveals the hyperconjugative energy about 19.04, 20.00, 20.05, 19.99, 20.10, 20.13, 19.41, 20.11, 19.96, and 20.30 kJ/mol for selected benzimidazole and 2-phenyl benzimidazoles respectively. This interaction could be revealed that the delocalization occurs in five membered ring for benzimidazole, due to the presence of C=N-C. Similarly the conjugative π bonds in the phenyl ring shows maximum delocalization during the interaction with π* acceptor bonds. It is evident from benzimidazoles that the πC2-C5, C3-C6 and C4-C7 delocalize more energy to the acceptor bond (π* acceptor). The electron density of donor bonds decreases while the acceptor (π*) bond electron density increases. Investigation of NBO analysis is described the stabilization energy for the conjugative interaction or charge transfer between the donor and acceptor bond orbitals [85, 86]. The interactions of π(C-C) with π*(C-C), π*(N-C) and π(N-C) with π*(C-C) are more responsible for the conjugation of respective π* bonds in benzimidazole and substituted 2-phenyl benzimidazole. The investigated molecules are divided into parts from the results of NBO analysis. One part is benzimidazole moiety and other benzene ring without and with substitutions. From the **Table 3** these conjugative interactions are formed with close stabilization energy in the range from 17.00 to 23.00 kcal/mol. The stabilization energy values for these interactions are agreed with literature values [87]. From the **Table 3**, the π*(N11-C12) delocalizes the maximum energy to π(C4-C7) and (C15-C18) bond respectively for all benzimidazoles in the range from 54.62 to 190.30 kJ/mol. Similarly, the π*(C18-C19) bond transfers the high energy about 247.57 kJ/mol to (C16-C20) bond for 2-(o-hydroxyphenyl)benzimidazole. The second order perturbation energies associated with the hyperconjugative interactions in NBO basis confirms the presence of intermolecular interactions.

4.4 Mulliken atomic charges

Thus the important quantum mechanical calculations further applied to calculate the atomic charges for molecular system [88]. The charge distributions of all atoms present in benzimidazole molecules are calculated by the Mulliken method [89]. The Mulliken atomic charges of selected benzimidazole molecules are presented in

DOI: http://dx.doi.org/10.5772/intechopen.101966

Donor (i)	Acceptor (j)	Benz imidazole	2-phenyl benz imidazole	2-(m-methyl phenyl) benz imidazole	2-(m-methoxy phenyl) benz imidazole	2-(p-methyl phenyl) benz imidazole	2-(p-methoxy phenyl) benz imidazole	2-(o-hydroxy phenyl) benz imidazole	2-(p-hydroxy phenyl) benz imidazole	2-(o-amino phenyl) benz imidazole	2-(p-amino phenyl) benz imidazole
		E(2) kJ mol^{-1}									
π(C2-C5)	π^*(C3-C6)	22.29	22.12	22.11	22.15	22.07	22.07	22.04	18.14	21.89	21.97
	π^*(C4-C7)	17.73	17.83	17.82	17.86	17.82	17.85	18.44	20.15	18.11	17.82
π(C3-C6)	π^*(C2-C5)	18.03	18.09	18.09	18.07	18.10	18.09	18.32	22.02	18.32	18.12
	π^*(C4-C7)	20.13	20.22	20.18	20.20	20.15	20.13	20.66	17.87	20.27	20.03
π(C4-C7)	π^*(C2-C5)	18.49	18.36	18.39	18.36	18.40	18.42	18.13	18.19	18.35	18.48
	π^*(C3-C6)	18.04	18.04	18.09	18.09	18.14	18.20	17.89	18.40	18.18	18.35
	π^*(N11-C12)	18.57	16.73	16.63	16.63	16.59	16.52	15.87	16.53	16.04	16.26
π(N11-C12)	π^*(C4-C7)	19.04	20.00	20.05	19.99	20.10	20.13	19.41	20.11	19.96	20.30
	π^*(C15-C18)		10.12	9.90	7.86	9.90	9.78	6.71			
π(C15-C18)	π^*(N11-C12)		20.60	19.97	19.34	21.02	20.85	24.94	21.75		21.30
	π^*(C16-C19)		19.80	21.11	20.93	18.45	17.83	15.71	18.02		17.24
	π^*(C17-C20)		19.16	18.70	19.77	19.77	20.33	22.74	23.49		21.91
π(C16-C19)	π^*(C15-C18)		20.48	19.66	19.59	22.02	20.88	23.93	22.49	22.59	22.66
	π^*(C17-C20)		19.27	20.46	22.97	18.20	16.32	19.18	17.80	18.17	16.42
π(C17-C20)	π^*(C15-C18)		20.55	21.16	20.07	19.66	18.76	16.50	17.22	16.41	17.22
	π^*(C16-C19)		21.84	20.49	17.04	23.62	23.38	21.24	22.95	22.36	23.69
LP(1)N11	π^*(C4-C7)	6.95						33.83	33.78		
	π^*(C12-N13)	8.28						52.81	49.09		
LP(1)N13	π^*(C4-C7)		33.96	33.96	33.82	33.91	33.79			34.03	33.76

Donor (i)	Acceptor (j)	Benz imidazole	2-phenyl benz imidazole	2-(m-methyl phenyl) benz imidazole	2-(m-methoxy phenyl) benz imidazole	2-(p-methyl phenyl) benz imidazole	2-(p-methoxy phenyl) benz imidazole	2-(o-hydroxy phenyl) benz imidazole	2-(p-hydroxy phenyl) benz imidazole	2-(o-amino phenyl) benz imidazole	2-(p-amino phenyl) benz imidazole
	π^*(N11-C12)		49.07	49.00	48.72	49.06	48.92			51.13	48.93
LP(1)O25	π^*(C18-C19)				31.79			34.82	30.14		
π^*(N11-C12)	π^*(C4-C7)	54.62		78.65	79.23	77.72	77.98	50.24	75.98	62.76	80.03
	π^*(C15-C18)		167.75	152.81	100.02	166.29	190.39	84.58	148.02	77.69	174.06
π^*(C18-C19)	π^*(C15-C17)				151.75						
	π^*(C16-C20)				166.50		168.46	247.57	272.37		

Table 3.
Second order perturbation theory analysis of Fock matrix in NBO basis for selected benzimidazoles by DFT B3LYP 6-31G method.

Atoms	Benz imidazole	2-phenyl benz imidazole	2-(m-methyl phenyl) benz imidazole	2-(m-methoxy phenyl) benz imidazole	2-(p-methyl phenyl) benz imidazole	2-(p-methoxy phenyl) benz imidazole	2-(o-hydroxy phenyl) benz imidazole	2-(p-hydroxy phenyl) benz imidazole	2-(o-amino phenyl) benz imidazole	2-(p-amino phenyl) benz imidazole
1H	0.138261	0.137100	0.136782	0.137655	0.136121	0.135958	0.141845	0.126152	0.134702	0.133829
2 C	−0.097181	−0.106860	−0.107260	−0.107084	−0.107632	−0.108005	−0.097640	−0.074372	−0.106858	−0.109524
3 C	−0.066378	−0.073803	−0.073798	−0.073607	−0.074250	−0.074668	−0.073868	−0.108222	−0.073166	−0.075092
4 C	0.032916	0.071322	0.070149	0.068002	0.071752	0.071510	0.110315	0.072285	0.079284	0.070580
5 C	−0.136317	−0.134176	−0.134219	−0.134114	−0.134392	−0.134414	−0.135183	−0.155754	−0.134554	−0.134574
6 C	−0.156456	−0.155168	−0.155464	−0.155447	−0.155462	−0.155708	−0.152607	−0.134226	−0.155256	−0.156420
7 C	0.3296018	0.3365298	0.335907	0.334665	0.336073	0.335654	0.324924	0.335776	0.333672	0.334464
8H	0.1229229	0.122071	0.121565	0.122217	0.121082	0.120684	0.126677	0.120736	0.122688	0.118882
9H	0.12254210	0.12186810	0.121294	0.121722	0.120797	0.120205	0.132516	0.120866	0.122601	0.118473
10H / N	0.12747411	0.12701911	0.126426	0.126032	0.126064	0.125161	−0.560586	0.135251	0.129138	0.123999
11 N / C	−0.40539712	−0.43623112	−0.435362	−0.431475	−0.437937	−0.438620	0.430471	−0.829967	−0.481336	−0.440923
12C/ N/H	0.25736613	0.35960913	0.355541	0.350613	0.358957	0.359337	−0.825723	0.321707	0.371033	0.357256
13 N/ H/C	−0.76618814	−0.82816914	−0.825954	−0.822124	−0.829186	−0.829467	0.330113	0.359544	−0.804057	−0.827406
14H/ C/N	0.32098015	0.32265315	0.322142	0.320084	0.321883	0.320376	0.187442	−0.441448	0.324986	0.319629
15H	0.175857									
15C		0.13689816	0.135817	0.125687	0.141260	0.137938	−0.118226	0.138396	0.124973	0.130048
16C		−0.11274417	−0.137633	−0.120574	0.106575	0.287632	−0.169922	0.258006	−0.114779	0.237503
17C		−0.14206118	−0.141589	−0.140008	−0.147270	−0.142365	0.174685	−0.154065	0.191570	−0.146137

Atoms	Benz imidazole	2-phenyl benz imidazole	2-(m-methyl phenyl) benz imidazole	2-(m-methoxy phenyl) benz imidazole	2-(p-methyl phenyl) benz imidazole	2-(p-methoxy phenyl) benz imidazole	2-(o-hydroxy phenyl) benz imidazole	2-(p-hydroxy phenyl) benz imidazole	2-(o-amino phenyl) benz imidazole	2-(p-amino phenyl) benz imidazole
18C		−0.14786719	−0.169322	−0.155106	−0.152066	−0.156250	−0.112405	−0.149417	−0.160741	−0.152654
19C		−0.13676620	0.090906	0.265929	−0.159726	−0.145257	−0.137129	−0.149145	−0.141445	−0.144521
20C / H		−0.13440521	−0.135466	−0.132738	−0.157269	−0.141035	0.147321	−0.119925	−0.144329	−0.143119
21H		0.17151722	0.167601	0.165297	0.169863	0.173190	0.123446	0.114675	0.121586	0.169114
22H		0.10943423	0.104834	0.115260	0.107799	0.111810	0.132189	0.175335	0.120184	0.109992
23H		0.12886324	0.130958	0.134836	0.124976	0.131191	0.127431	0.128766	0.122846	0.117131
24H		0.13342825	0.124760	0.141672	0.129645	0.145411	0.110820	0.144332	0.126761	0.122272
25H/ C/O/N		0.129940	−0.481792	−0.546460	−0.471062	−0.541636	−0.613107	−0.607400	−0.721965	−0.721135
26H/C			0.161815	−0.172963	0.159713	−0.175896	0.396196	0.372113	0.284914	0.293247
27H			0.148639	0.149400	0.147909	0.149136			0.327550	0.295088
28H			0.142723	0.170147	0.145782	0.169412				
29H				0.142482		0.148717				

Table 4.
The computed Mulliken atomic charges for selected benzimidazole molecules given using DFT B3LYP 6-31G method.

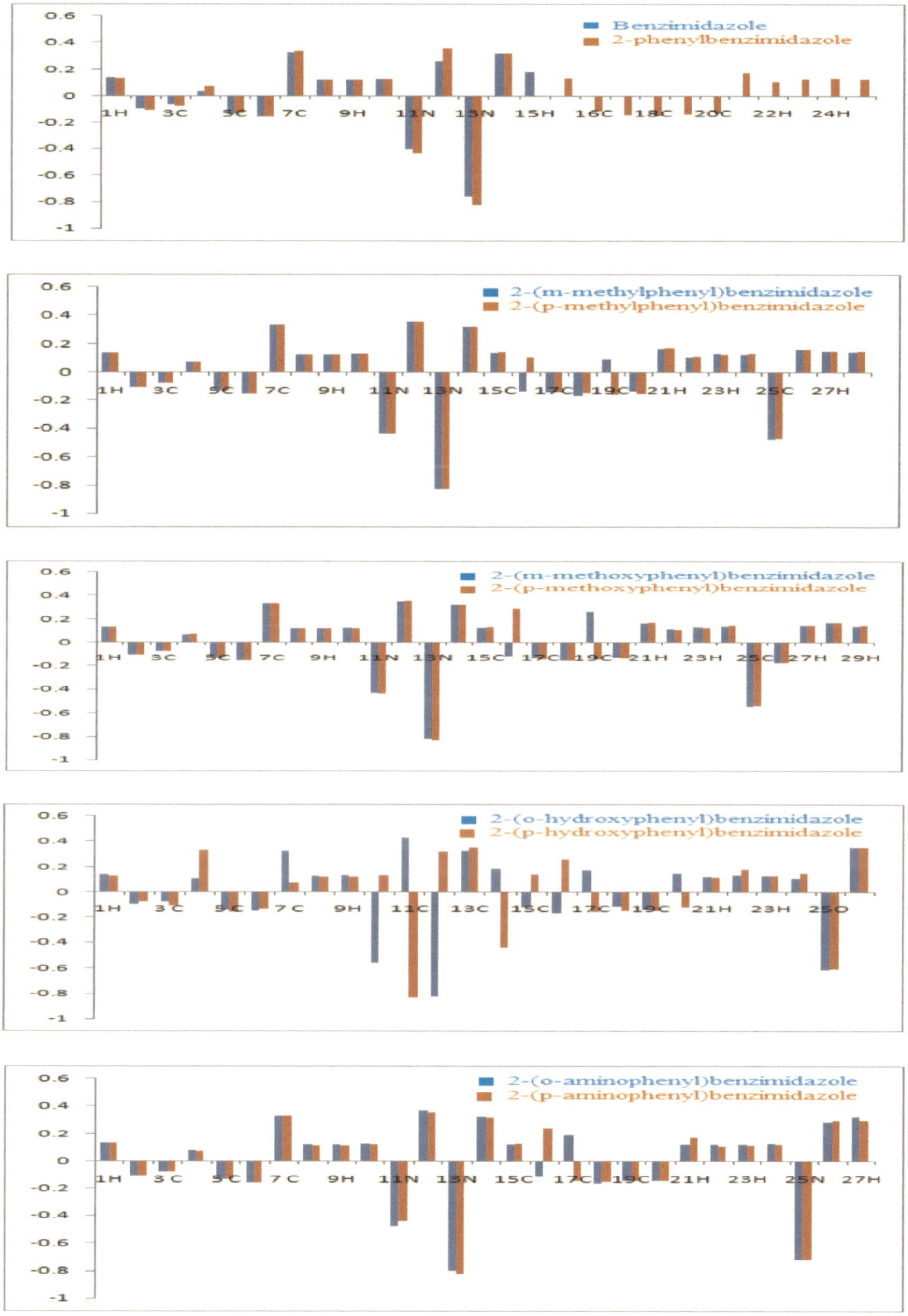

Figure 5.
The computed Mulliken charges for all atoms in selected benzimidazoles.

Table 4 and shown in **Figure 5**. The Mulliken atomic charges were computed at the DFT B3LYP 6-31G method. The carbon atoms numbering C4, C7 and C12 are shown with the positive values except other carbon atoms in the whole system. These results expected that these carbon atoms are connected with electronegative nitrogen atoms in benzimidazole [90]. In benzimidazole, C7 atom is bonded with N13-H having high positive value (0.329 a.u.) and C4 atom with N11 atom having less positive value (0.032 a.u.). The other benzimidazole molecules are following the same trend. A positive charge of all the hydrogen atoms are displayed in **Table 4**, but H26 was gained maximum positive charge than the other hydrogen atoms, due to the presence of electronegative atom (O25) in o-hydroxy and p-hydroxyphenyl benzimidazole when compared with the hydrogen in amino group lesser values. The presence of three nitrogen atoms in o-amino and p-aminophenyl benzimidazole (N11 = −0.481 a.u., N13 = −0.804 a.u. and N25 = −0.721 a.u.) are shown in different environment because N13 atom more negative values.

5. Conclusion

Ten compounds of benzimidazole and 2-phenyl substituted benzimidazoles such as (1) benzimidazole, (2) 2-phenylbenzimidazole, (3) 2-(m-methylphenyl)benzimidazole, (4) 2-(p-methylphenyl)benzimidazole, (5) 2-(m-methoxyphenyl)benzimidazole, (6) 2-(p-methoxyphenyl)benzimidazole, (7) 2-(o-hydroxyphenyl) benzimidazole, (8) 2-(p-hydroxyphenyl)benzimidazole, (9) 2-(o-aminophenyl)benzimidazole and (10) 2-(p-aminophenyl)benzimidazole were selected to study for the spectral and theoretical properties. Synthesis of these molecules by many methods were discussed and given the reaction scheme. Then the absorption and fluorescence spectrum of all molecules were given with changing the wavelength respect to the substitution of groups in the benzene ring. The conjugation and maximum also involved and π-π^* transition possible in the absorption spectrum. Further, DFT method was used to determine the structural parameters, energy values, HOMO-LUMO energy gap, thermodynamic parameters, molecular electrostatic potential map for all molecules. NBO analysis was revealed the hyperconjugative interaction between bonding orbitals, lone pair orbitals and antibonding orbitals and also calculated the stabilization energy of selected bonded orbitals. Finally the computed Mulliken atomic charges was determined using DFT B3LYP 6-31G method.

Author details

A. Antony Muthu Prabhu
Department of PG Chemistry, Aditanar College of Arts and Science, Tiruchendur, Tamilnadu, India

*Address all correspondence to: antonyphdchem@yahoo.com

References

[1] Tonelli M, Simone M, Tasso B, Novelli F, Boido V, Sparatore F, et al. Activiral activity of benzimidazole derivatives. II. Activiral activity of 2-phenylbenzimidazole derivatives. Bioorganic & Medicinal Chemistry. 2010;**18**:2937-2953. DOI: 10.1016/j.bmc.2010.02.037

[2] Mann J, Baron A, Opoku-Boahen Y, Johansson E, Parkinson G, Kelland LR, et al. A new class of symmetric bisbenzimidazole-based DNA minor groove-binding agents showing antitumor activity. Journal of Medicinal Chemistry. 2001;**44**:138-144. DOI: 10.1021/jm000297b

[3] Iemura R, Hori M, Saito T, Ohtaka H. Bioisosteric transformation of H1-antihistaminic benzimidazole derivatives. Chemical & Pharmaceutical Bulletin. 1989;**37**:2723-2726. DOI: 10.1248/cpb.37.2723

[4] Ansari KF, Lal C. Synthesis, physicochemical properties and antimicrobial activity of some new benzimidazole derivatives. European Journal of Medicinal Chemistry. 2009;**44**:4028-4033. DOI: 10.1016/j.ejmech.2009.04.037

[5] Preston PN. Synthesis, reactions and spectroscopic properties of benzimidazoles. Chemical Reviews. 1974;**74**:279-314. DOI: 10.1021/cr60289a001

[6] Márquez-Navarro A, Nogueda-Torresa B, Hernández-Campos A, Soria-Artechec O, Castillo R, Rodríguez-Morales S, et al. Anthelmintic activity of benzimidazole derivatives against *Toxocara canis* second-stage larvae and *Hymenolepis nana* adults. Acta Tropica. 2009;**109**:232-235. DOI: 10.1016/j.actatropica.2008.11.014

[7] Refaat HM. Synthesis and anticancer activity of some novel 2-substituted benzimidazole derivatives. European Journal of Medicinal Chemistry. 2010;**45**:2949-2956. DOI: 10.1016/j.ejmech.2010.03.022

[8] Sharma S, Gangal S, Rauf A. Convenient one-pot synthesis of novel 2-substitutedbenzimidazoles, tetrahydrobenzimidazoles and imidazoles and evaluation oftheir in vitro antibacterial and antifungal activities. European Journal of Medicinal Chemistry. 2009;**44**:1751-1757. DOI: 10.1016/j.ejmech.2008.03.026

[9] Zhang D, Wang Z, Xu W, Sun F, Tang L, Wang J. Design, synthesis and antibacterial activity of novel actinonin derivatives containing benzimidazole heterocycles. European Journal of Medicinal Chemistry. 2009;**44**:2202-2210. DOI: 10.1016/j.ejmech.2008.05.009

[10] Kaplancikli ZA, Turan-Zitouni G, Ozdemir A, Can Ozgur D, Chevallet P. Synthesis and Antinociceptive Activities of Some Pyrazoline Derivatives. European Journal of Medicinal Chemistry. 2009;**44**:2606-2610. DOI: 10.1007/s00044-014-1309-1

[11] Sahoo BM, Banik BK, Mazaharunnisa RNS, Raju B. Microwave assisted green synthesis of benzimidazole derivatives and evaluation of their anticonvulsant activity. Current Microwave Chemistry. 2019;**6**:23-29. DOI: 10.2174/2213335606666190429124745

[12] Patil A, Ganguli S, Surana S. Synthesis And Antiulcer Activity Of 2-[5-Substituted-1-H-Benzo(D) Imidazole-2-Yl Sulfinyl] Methyl-3-Substituted Quinqzoline-4-(3H)Ones.

Journal of Chemical Sciences. 2010; **3**(122):443-449. DOI: 10.1007/s12039-010-0052-5

[13] Aljourani J, Golozar MA, Raessi K. The inhibition of carbon steel corrosion in hydrochloric and sulfuric acid media using some benzimidazole derivatives. Materials Chemistry and Physics. 2010; **121**:320-325. DOI: 10.1016/j.matchemphys.2010.01.040

[14] Obot IB, Obi-Egbedi NO. Theoretical study of benzimidazole and its derivatives and their potential activity as corrosion inhibitors. Corrosion Science. 2010;**52**:657-660. DOI: 10.1016/j.corsci.2009.10.017

[15] Dutta A, Saha S, Adhikari U, Banerjee P, Sukul D. Effect of substitution on corrosion inhibition properties of 2-(substituted phenyl) benzimidazole derivatives on mild steel in 1M HCl solution: A combined experimental and theoretical approach. Corrosion Science. 2017;**123**:256-266. DOI: 10.1016/j.corsci.2017.04.017

[16] Kohler P. The biochemical basis of anthelmintic action and resistance. International Journal for Parasitology. 2001;**31**:336-345. DOI: 10.1016/s0020-7519(01)00131-x

[17] Jubie S, Rajeshkumar R, Yella Reddy B, Siddhartha G, Sandeep M, Surendrareddy K, et al. Microwave assisted synthesis of some novel Benzimidazole submitted fluoroquinolones and their antimicrobial evaluation. Journal of Pharmaceutical Sciences and Research. 2010;**2**:69-76

[18] Bai Y, Lv X, Liu Z, Guo T, Li J. Synthesis, spectral and DFT study of 5-butyl-3,9-difluoro-5*H*-benzimidazo[1,2-*a*] Benzimidazole. Journal of Molecular Structure. 2018;**115**:111-125. DOI: 10.1016/j.molstruc.2017.08.102

[19] Akkoç S, Tüzün B, İlhan IO, Akkurt M. Investigation of structural, spectral, electronic, and biological properties of 1,3-disubstituted benzimidazole derivatives. Journal of Molecular Structure. 2020;**1219**:128582-128592. DOI: 10.1016/j.molstruc.2020.128593

[20] Nour T, Ghani A, Mansour AM. Molecular structure of 2-chloromethyl-1H-benzimidazole hydrochloride: Single crystal, spectral, biological studies, and DFT calculations. Spectrochimica Acta A. 2012;**86**:605-613. DOI: 10.1016/j.saa.2011.11.024

[21] Li GY, Geng JC, He CH, Cui GH. Structure, spectroscopy, and theory calculations on 1,3-bis((5,6-dimethyl-1H-benzo[d]imidazol-1-yl)methyl) benzene. Journal of Molecular Structure. 2013;**1031**:56-60. DOI: 10.1016/j.molstruc.2012.07.014

[22] Patil VS, Padalkar VS, Tathe AB, Gupta VD, Sekar N. Synthesis, Photo-physical and DFT Studies of ESIPT Inspired Novel 2-(2′,4′-Dihydroxyphenyl) Benzimidazole, Benzoxazole and Benzothiazole. Journal of Fluorescence. 2013;**23**:1019-1029. DOI: 10.1007/s10895-013-1228-4

[23] Sinha H, Dogra S. Ground and excited state prototropic reactions in 2-(o-hydroxyphenyl) benzimidazole. Chemical Physics. 1986;**102**:337-347. DOI: 10.1016/0301-0104(86)80006-4

[24] Douhal A, Amat-Guerri F, Lillo M, Acuna A. Proton transfer spectroscopy of 2-(2′-hydroxyphenyl) imidazole and 2-(2′-hydroxyphenyl) benzimidazole dyes. Journal of Photochemistry and Photobiology A: Chemistry. 1994;**78**:127-138. DOI: 10.1016/1010-6030(93)03724-U

[25] Padalkar VS, Tathe AB, Gupta VD, Patil VS, Phatangare KR, Sekar N.

Synthesis and photo-physical characteristics of ESIPT inspired 2-substituted benzimidazole, benzoxazole and benzothiazole fluorescent derivatives. Journal of Fluorescence. 2012;**22**:311-322. DOI: 10.1007/s10895-011-0962-8

[26] Sandra S, Dogra S. Excited state intramolecular protons transfer in 2-(2′-N-palmitoyl-aminophenyl) benzimidazole: Effect of carbonyl group. Journal of Molecular Structure. 1999; **476**:223-233. DOI: 10.1016/S0022-2860(98)00550-X

[27] Manojai N, Daengngern R, Kerdpol K, Ngaojampa C, Kungwan N. Heteroatom effect on photophysical properties of 2-(2′-hydroxyphenyl) benzimidazole and its derivatives as fluorescent dyes: A TD-DFT study. Journal of Luminescence. 2017;**188**: 275-282. DOI: 10.1016/j.jlumin.2017.04.040

[28] Al-Jorani KR, Abbas RF, Waheb AA. Synthesis and characterization of a novel benzimidazole derivative: Solvatochromic and acid-base indicator applications. Materials Today: Proceedings. 2021;**42**:1791-1799. DOI: 10.1016/j.matpr.2020.12.184

[29] Vazquez SR, Rodriguez MCR, Mosquera M. Rotamerism, tautomerism, and excited-state intramolecular proton transfer in 2-(4′-N, N-diethylamino-2′-hydroxyphenyl)benzimidazoles: Novel benzimidazoles undergoing excited-state intramolecular coupled proton and charge transfer. The Journal of Physical Chemistry. A. 2008;**112**:376-387. DOI: 10.1021/jp076634a

[30] Henary MM, Wu Y, John Cody S, Sumalekshmy S, Li J, Mandal S, et al. Excited-state intramolecular proton transfer in 2-(2’-Arylsulfonamidophenyl)benzimidazole derivatives: The effect of donor and acceptor substituents. The Journal of Organic Chemistry. 2007;**72**:4784-4797. DOI: 10.1021/jo070433l

[31] Santra S, Dogra SK. Excited-state intramolecular proton transfer in 2-(2′-Aminophenyl)benzimidazole. Chemical Physics. 1998;**226**:285-296. DOI: 10.1016/S0301-0104(97)00311-X

[32] Chipem FAS, Behera SK, Krishnamoorthy G. Enhancing excited state intramolecular proton transfer in 2-(2′-Hydroxyphenyl)benzimidazole and its nitrogen-substituted analogues by β-Cyclodextrin: The effect of nitrogen substitution. The Journal of Physical Chemistry. A. 2013;**117**:4084-4095. DOI: 10.1021/jp311438s

[33] Golcuk K, Altun A, Kumru M. Thermal studies and vibrational analyses of m-methylaniline complexes of Zn(II), Cd(II) and Hg(II)bromides. Spectrochimica Acta A. 2003;**59**: 1841-1847. DOI: 10.1016/S1386-1425(02)00415-8

[34] Vaschetto ME, Retamal BA, Monkman AP. Density functional studies of aniline and substituted anilines. Journal of Molecular Structure: THEOCHEM. 1999;**468**:209-221. DOI: 10.1016/S0166-1280(98)00624-1

[35] Clark T, Chandrasekhar J, Spitznagel GW, Schleyer PVR. Efficient diffuse function-augmented basis sets for anion calculations. III. The 3-21+G basis set for first-row elements, Li-F. Journal of Computational Chemistry. 1983;**4**:294-301. DOI: 10.1002/jcc.540040303

[36] Forsyth DA, Sebag AB. Computed ^{13}C NMR chemical shifts via empirically scaled GIAO shielding and molecular mechanics geometries. Conformation and configuration from ^{13}C shifts. Journal of the American Chemical

Society. 1997;**119**:9483-9494. DOI: 10.1021/ja970112z

[37] Gavin Tsai HH, Sara Sun HH, Tan CJ. TD-DFT Study of the Excited-State Potential Energy Surfaces of 2-(2′-Hydroxyphenyl)benzimidazole and its amino derivatives. The Journal of Physical Chemistry. A. 2010;**114**: 4065-4079. DOI: 10.1021/jp100022y

[38] Chipem FAS, Dash N, Krishnamoorthy G. Role of nitrogen substitution in phenyl ring on excited state intramolecular proton transfer and rotamerism of 2-(2′-Hydroxyphenyl)-benzimidazole: A theoretical study. The Journal of Chemical Physics. 2011;**134**: 1-9. DOI: 10.1063/1.3562124

[39] Prommin C, Kanlayakan N, Chansen W, Salaeh R, Kerdpol K, Daengngern R, et al. Theoretical insights on solvent control of intramolecular and intermolecular proton transfer of 2-(2'-Hydroxyphenyl)benzimidazole. The Journal of Physical Chemistry. A. 2017;**121**: 5773-5784. DOI: 10.1021/acs.jpca.7b03454

[40] Oturak H, Kınaytürk NK, Çırak C. Experimental and theoretical spectral (FT-IR, Raman, NMR, UV-Vis and NLO) analysis of a potential antitumor drug: 1-Methyl-6-Nitro-1H-benzimidazole. Spectroscopy and Spectal Analysis. 2018; **38**:1963-1969. DOI: 10.3964/j.ssn.1000-0593(2018)06-1963-07

[41] Karnan M, Balachandran V, Murugan M, Murali MK. Quantum chemical vibrational study, molecular property, FTIR, FT-Raman spectra, NBO, HOMO-LUMO energies and thermodynamic properties of 1-methyl-2-phenyl benzimidazole. Spectrochimica Acta A. 2014;**130**:143-151. DOI: 10.1016/j.saa.2014.03.128

[42] Mary YS, Jojo PJ, Panicker CY, Alsenoy CV, Ataei S, Yildiz I. Theoretical investigations on the molecular structure, vibrational spectra, HOMO-LUMO and NBO analysis of 5-chloro-2-((4-chlorophenoxy) methyl) benzimidazole. Spectrochimica Acta A. 2014;**122**:499-511. DOI: 10.1016/j.saa.2013.11.025

[43] Sundaraganesan N, Ilakiamani S, Subramani P, Joshua BD. Comparison of experimental and ab initio HF and DFT vibrational spectra of benzimidazole. Spectrochimica Acta A. 2007;**67**:628-635. DOI: 10.1016/j.saa.2006.08.020

[44] Sekerci M, Atalay Y, Yakuphanoğlu F, Başoğlu DA. A theoretical study on 1-(thiophen-2-yl-methyl)-2-(thiophen-2-yl)-1H-benzimidazole. Spectrochimica Acta A. 2007;**67**:503-508. DOI: 10.1016/j.saa.2006.08.007

[45] Xavier TS, Rashid N, Joe IH. Vibrational spectra and DFT study of anticancer active molecule 2-(4-Bromophenyl)-1H-benzimidazole by normal coordinate analysis. Spectrochimica Acta A. 2011;**78**:319-326. DOI: 10.1016/j.saa.2010.10.013

[46] Foujji LE, Bourakadi KE, Essassi EM, Boeré RT, Qaiss A, Bouhfid R. Solid-state zwitterionic tautomerization of 2-((5-methyl-1*H*-pyrazol-3-yl)methyl)-1*H*-benzimidazole: Synthesis, characterization, DFT calculation and docking studies. Journal of Molecular Structure. 2021;**1235**:130231-130241. DOI: 10.1016/j.molstruc.2021.130231

[47] Nour T, Ghani A, Mansour AM. Molecular structures of 2-arylaminomethyl-1H-benzimidazole: Spectral, electrochemical, DFT and biological studies. Spectrochimica Acta A. 2012;**91**:272-284. DOI: 10.1016/j.saa.2012.01.080

[48] Unal A, Eren B. FT-IR, dispersive Raman, NMR, DFT and antimicrobial

activity studies on 2-(Thiophen-2-yl)-1H-benzo[d]imidazole. Spectrochimica Acta A. 2013;**114**:129-136. DOI: 10.1016/j.saa.2013.05.045

[49] Nour T, Ghani A, Mansour AM. 2-[(1H-Benzimidazol-2-ylmethyl)-amino]-benzoic acid methyl ester: Crystal structure, DFT calculations and biological activity evaluation. Spectrochimica Acta A. 2011;**81**:754-763. DOI: 10.1016/j.saa.2011.07.021

[50] Furniss BS, Hannaford AJ, Smith PWG, Tatchell AR, Vogel's. Textbook of Practical Organic Chemistry. 5th ed. New York: John Wiley & Sons; 1989. p. 1162. DOI: 10.1002/pola.1991.080290821

[51] El-Gohary NS, Shaaban MI. Synthesis and biological evaluation of a new series of benzimidazole derivatives as antimicrobial, antiquorum-sensing and antitumor agents. European Journal of Medicinal Chemistry. 2017;**131**:255-262. DOI: 10.1016/j.ejmech.2017.03.018

[52] Adegboye AA, Khan KM, Salar U, Aboaba SA, Kanwal CS, Fatima I, et al. 2-Aryl benzimidazoles: Synthesis, In vitro α-amylase inhibitory activity, and molecular docking study. European Journal of Medicinal Chemistry. 2018; **150**:248-260. DOI: 10.1016/j.ejmech.2018.03.011

[53] Nannapaneni DT, Gupta Atyam VSSS, Reddy MI, Sarva Raidu C. Synthesis, characterization, and biological evaluation of benzimidazole derivatives as potential anxiolytics. Journal of Young Pharmacists. 2010; **2**(3):273-279. DOI: 10.4103/0975-1483.66809

[54] Patil SV, Patil SS, Bobade VD. A simple and efficient approach to the synthesis of 2-substituted benzimidazole via sp^3 C–H functionalization. Arabian Journal of Chemistry. 2016;**9**:S515-S521. DOI: 10.1016/j.arabjc.2011.06.017

[55] Hoffman K. The Chemistry of Heterocyclic Compounds. Imidazole and Its Derivatives, Part I, Interscience. New York: Wiley; 1951

[56] Tway PC, Cline Love LJ. Photophysical properties of benzimidazole and thiabendazole and their homologues. Effect of substituents and solvent on the nature of the transition. The Journal of Physical Chemistry. 1982; **86**:5223-5226. DOI: 10.1021/j100223a031

[57] Mishra AK, Dogra SK. Effect of solvents and pH on the absorption and fluorescence spectra of 2-phenylbenzimidazole. Spectrochimica Acta A. 1983;**39**:609-611. DOI: 10.1016/0584-8539(83)80033-6

[58] Roberts EL, Dey J, Warner IM. Ground- and excited-state structural orientation of 2-(2'-Hydroxyphenyl) benzazoles in cyclodextrins. The Journal of Physical Chemistry. 1996;**100**: 19681-19686. DOI: 10.1021/jp961500a

[59] Roberts EL, Dey J, Warner IM. Excited-state intramolecular proton transfer of 2-(2'-hydroxyphenyl) benzimidazole in cyclodextrins and binary solvent mixtures. The Journal of Physical Chemistry. A. 1997;**101**: 5296-5301. DOI: 10.1021/jp971232w

[60] Fores M, Duran M, Sola M, Orozco M, Luque FJ. Theoretical evaluation of solvent effects on the conformational and tautomeric equilibria of 2-(2'-Hydroxyphenyl) benzimidazole and on its absorption and fluorescence spectra. The Journal of Physical Chemistry. A. 1999;**103**: 4525-4532. DOI: 10.1021/jp984773+

[61] Chipem FAS, Krishnamoorthy G. Comparative theoretical study of

rotamerism and excited state intramolecular proton transfer of 2-(2′-Hydroxyphenyl)benzimidazole, 2-(2′-hydroxyphenyl)imidazo[4,5-b]pyridine, 2-(2′-Hydroxyphenyl)imidazo[4,5-c] pyridine and 8-(2′-Hydroxyphenyl) purine. The Journal of Physical Chemistry. A. 2009;**113**:12063-12070. DOI: 10.1021/jp903535g

[62] Sahu S, Das M, Krishnamoorthy G. Switching between CIS and trans anions of 2-(2′-hydroxyphenyl)benzimidazole: A molecular rotation perturbed by chemical stabilization. Physical Chemistry Chemical Physics. 2016;**18**: 11081-11090. DOI: 10.1039/ C5CP07949G

[63] Vázquez SR, Lustres JLP, Rodríguez-Prieto F, Mosquera M, Ríos Rodríguez MC. Excited-state proton and charge transfer in protonated amino and methylated derivatives of 2-(2′-hydroxyphenyl)benzimidazole. The Journal of Physical Chemistry. B. 2015; **119**:2475-2489. DOI: 10.1021/jp507917u

[64] Behera SK, Sadhuragiri G, Elumalai P, Sathiyendiran M, Krishnamoorthy G. Exclusive excited state intramolecular proton transfer from a 2-(2′-hydroxyphenyl) benzimidazole derivative. RSC Advances. 2016;**6**:59708-59717. DOI: 10.1039/C6RA11780E

[65] Akutsu K, Mori S, Shinmei K, Iwase H, Nakano Y, Fujii Y. Investigation of substitution effect on fluorescence properties of Zn^{2+}-selective ratiometric fluorescent compounds: 2-(2′-Hydroxyphenyl)benzimidazole derivatives. Talanta. 2016;**146**:575-584. DOI: 10.1016/j.talanta.2015.09.001

[66] Mishra AK, Dogra SK. Photoluminescence of 2-(o-aminophenyl)benzimidazole. Journal of Photochemistry. 1985;**31**: 333-344. DOI: 10.1016/0047-2670(85) 85101-7

[67] Mishra AK, Dogra SK. Effects of solvents and pH on the spectral behavior of 2-(p-aminophenyl)benzimidazole. Bulletin of the Chemical Society of Japan. 1985;**58**:3587-3592. DOI: 10.1246/ bcsj.58.3587

[68] Fleming I. Frontier Orbitals and Organic Chemical Reactions. New York: John Wiley & Sons; 1976

[69] Mansour AM. Molecular structure and spectroscopic properties of novel manganese(II)complex with sulfamethazine drug. Journal of Molecular Structure. 2013;**1035**: 114-123. DOI: 10.1016/j. molstruc.2012.09.048

[70] Nora M, Fatiha M, Leila N, Sakina H, Eddine KD. Density functional study of inclusion complex of albendazole/cucurbit [7]uril: Structure, electronic properties, NBO, GIAO and TD-DFT analysis. Journal of Molecular Liquids. 2015;**211**:40-47. DOI: 10.1016/j.molliq.2015.06.054

[71] Alyar S, Tülin S. Synthesis, spectroscopic characterizations, enzyme inhibition , molecular docking study and DFT calculations of new Schiff bases of sulfa drugs. Journal of Molecular Structure. 2019;**1185**:416-424. DOI: 10.1016/j.molstruc.2019.03.002

[72] Mondal S, Mandal SM, Mondal TK, Sinha C. Structural characterization of new Schiff bases of sulfamethoxazole and sulfathiazole, their antibacterial activity and docking computation with DHPS protein structure. Spectrochimica Acta A. 2015;**150**:268-279. DOI: 10.1016/ j.saa.2015.05.049

[73] Banuppriya G, Sribalan R, Padmini V. Synthesis and characterization of curcumin-sulfonamide hybrids: Biological

evaluation and molecular docking studies. Journal of Molecular Structure. 2018;**1155**: 90-100. DOI: 10.1016/j.molstruc.2017.10.097

[74] Glendening ED, Reed AE, Carpenter JE, Weinhold F. NBO Version 3.1. Pittsburgh: Gaussian Inc.; 2003

[75] Antony Muthu Prabhu A, Suresh Kumar GS, Fatiha M, Sorimuthu S, Sundar RM. Encapsulation of phenylalanine and 3,4-dihydroxyphenylalanine into β-cyclodextrin: Spectral and molecular modeling studies. Journal of Molecular Structure. 2015;**1079**:370-382. DOI: 10.1016/j.molstruc.2014.08.045

[76] Antony Muthu Prabhu A, Fatiha M, Leila N, Anantha Raj T, Navarro-Gonzalez I, Periago MJ, et al. Investigation of 3D contour map and intermolecular interaction of dopamine with β-Cyclodextrin and 2-hydroxypropyl-β-cyclodextrin. Journal of Solution Chemistry. 2018;**47**:409-429. DOI: 10.1007/s10953-018-0728-x

[77] Foster AJ, Weinhold F. Natural hybrid orbitals. Journal of the American Chemical Society. 1980;**102**:7211-7218

[78] Glendening ED, Badenhoop JK, Reed AE, Carpenter JE, Bohmann JA, Morales CM, et al. NBO Version 3.1. Madison: University of Wisconsin; 1998. p. 65

[79] Mariappan G, Sundaraganesan N, Manoharan S. Experimental and theoretical spectroscopic studies of anticancer drug rosmarinic acid using HF and density functional theory. Spectrochimica Acta A. 2012;**97**:340-351. DOI: 10.1016/j.saa.2012.06.011

[80] Balachandran V, Janaki A, Nataraj A. Theoretical investigations on molecular structure, vibrational spectra, HOMO, LUMO, NBO analysis and hyperpolarizability calculations of thiophene-2-carbohydrazide. Spectrochimica Acta A. 2014;**118**: 321-330. DOI: 10.1016/j.saa.2013.08.091

[81] Szafran M, Komasa A, Bartoszak-Adamska E. Crytsal and molecular structure of 4-carboxypiperidinium chloride(4-piperidinecarboxylic acid hydrochloride). Journal of Molecular Structure. 2007;**827**:101-107. DOI: 10.1016/j.molstruc.2006.05.012

[82] Gangadharan RP. Natural bond orbital (NBO) population analysis of 1-Azanapthalene-8-ol. Acta Physica Polonica A. 2014;**125**:18-22. DOI: 10.12693/APhysPolA.125.18

[83] Bhuvaneswari R, Bharathi MD, Anbalagan G, Chakkaravarthi G, Murugesan KS. Molecular structure, vibrational spectroscopic (FT-IR, FTRaman), NBO, HOMO and LUMO analysis of morpholinium oxalate by density functional method. Journal of Molecular Structure. 2018;**1173**: 188-195. DOI: 10.1016/j.molstruc.2018.06.109

[84] Khalil A, Abdel S, Kh A. TD-DFT calculations, NBO analysis and electronic absorption spectra of some thiazolo [3,2-a] pyridine derivatives. Journal of Molecular Structure. 2017;**1147**:651-667. DOI: 10.1016/j.molstruc.2017.06.098

[85] Ravikumar C, Huber Joe I, Jayakumar VS. Charge transfer interactions and nonlinear optical properties of push-pull chromophore benzaldehyde phenylhydrazone: A vibrational approach. Chemical Physics Letters. 2008;**460**:552-558. DOI: 10.1016/j.cplett.2008.06.047

[86] Sundius T. Scaling of ab initio force fields by MOLVIB. Vibrational

Spectroscopy. 2002;**29**:89-95. DOI: 10.1016/S0924-2031(01)00189-8

[87] Mumit MA, Pal TK, Alam MA, Islam MAA, Paul S, Sheikh MC. DFT studies on vibrational and electronic spectra, HOMO-LUMO, MEP, HOMA, NBO and molecular docking analysis of benzyl-3-N-(2,4,5-trimethoxyphenylmethylene)hydrazine carbodithioate. Journal of Molecular Structure. 2020;**1220**:128715-128726. DOI: 10.1016/j.molstruc.2020.128715

[88] Gunasekaran S, Kumaresan S, Arunbalaji R, Anand G, Srinivasan S. Density functional theory study of vibrational spectra and assignment of fundamental modes of dacarbazine. Journal of Chemical Sciences. 2008;**120**: 315-324. DOI: 10.1007/s12039-008-0054-8

[89] Mulliken RS. Electronic population analysis on LCAO-MO molecular wave functions. The Journal of Chemical Physics. 1955;**23**:1833-1840. DOI: 10.1063/1.1740588

[90] Mathammal R, Monisha NR, Yasaswini S, Krishnakumar V. Molecular structure, vibrational analysis (FT-IR, FT-Raman), NMR, UV, NBO and HOMO–LUMO analysis of N,N-Diphenyl Formamide based on DFT calculations. Spectrochimica Acta A. 2015;**139**:521-532. DOI: 10.1016/j.saa.2014.11.057